CASTLES
THEIR CONSTRUCTION AND HISTORY

SIDNEY TOY

SENIOR HIGH
Red Lion Area

DOVER PUBLICATIONS, INC.
NEW YORK

21962

To
MY WIFE
IN GRATITUDE FOR
CONSISTENT HELP AND ENCOURAGEMENT
THE BOOK
IS AFFECTIONATELY DEDICATED

Title-page illustration by A. G. Smith from *Cut & Assemble a Medieval Castle: A Full-Color Model of Caernarvon Castle in Wales* (Dover 24663-9); copyright © 1984 by A. G. Smith.

Published in Canada by General Publishing Company, Ltd., 30 Lesmill Road, Don Mills, Toronto, Ontario.
Published in the United Kingdom by Constable and Company, Ltd.

This Dover edition, first published in 1985, is an unabridged republication of the work originally published by William Heinemann Ltd, London, in 1939 under the title *Castles: A Short History of Fortifications from 1600* B.C. *to* A.D. *1600*. A number of typographic errors have been tacitly corrected.

Manufactured in the United States of America
Dover Publications, Inc., 31 East 2nd Street, Mineola, N.Y. 11501

Library of Congress Cataloging in Publication Data

Toy, Sidney.
   Castles : their construction and history.

   Reprint. Originally published: London : W. Heinemann, 1939.
   Includes index.
   1. Fortification—Europe—History. 2. Fortification—Near East—History. 3. Castles—Europe—History. 4. Castles—Near East—History. I. Title.
UG428.T69   1985   355.7′094   85-4598
ISBN 0-486-24898-4 (pbk.)

CONTENTS

CHAPTER		PAGE
	Preface	xiii
I.	The Ancient Fortresses of Babylon, Mesopotamia, Assyria, Asia Minor and Greece	1
II.	Fortifications of Greece and Rome, 300 B.C. to 200 B.C.	16
III.	Fortifications of the Levant, Rome, and Western Europe, 200 B.C. to 30 B.C.	21
IV.	Fortifications of the Roman Empire	29
V.	Byzantine Fortifications from the Fifth to the Tenth Century	41
VI.	Fortifications of Western Europe from the Fifth to the Eleventh Century	50
VII.	Rectangular Keeps or Donjons	66
VIII.	Byzantine and Saracen Fortifications of the Twelfth Century	83
IX.	Transitional Keeps of the Twelfth Century	90
X.	Fortifications and Buildings of the Bailey in the Eleventh and Twelfth Centuries	100
XI.	Castles from about 1190 to 1270	116
XII.	Siege Engines and Siege Operations of the Middle Ages	141
XIII.	Edwardian and Contemporary Castles	153
XIV.	Towns, Fortified Bridges, and Towers	170
XV.	Gatehouses and Defences of the Curtain during the Thirteenth and Fourteenth Centuries	189
XVI.	Development of the Tower-House	210
XVII.	Sixteenth-Century Forts	230
	Index	237

ILLUSTRATIONS

		PAGE
Aigues Mortes	Plans, Section and Elevation of the Tour de Constance	129
Anadoli Kavak	Plan of Hieron Castle	86
	The Gateway	124
Angers	The Château from the South	135
Arques	The Château from the South	64
Arundel Castle	Plan of the Castle	53
Ashby-de-la-Zouche	The Chapel	228
Athens	The Dipylon	12
Basement Windows		113
Beaumaris Castle	Plan of the Castle	162
	The North Gatehouse from within the Bailey	137
Beeston Castle	The Castle from the West	132
Bodiam Castle	Plan of the Castle	217
	The Castle from the North	220
	The South and West Ranges of the Courtyard	220
Borcovicium	Plan of the Roman Camp	37
Borthwick Castle	Plan of the Castle	221
	The Tower from the North-West	222
Bothwell Castle	The Keep from within the Bailey	137
Brinklow Castle	Plan of the Castle	53
Caerlaverock Castle	The Gatehouse	168
Caernarvon Castle	Plan of the Castle	158
	The Castle from the South-East	152
	The Eagle Tower from within the Bailey	167
	Plan and Section of the King's Gate	196
	The King's Gate from the North-East	200
Caerphilly Castle	Plan of the Castle	154
	The Screen Wall from the South-East	124
	Elevation and Section of the Outer Gateway	192
	The East Gateway from the Inner Bailey	188
	Details of the East Gatehouse	193
	Towers of the Screen Wall	211
Cahors	The Pont Valentré	173
Cairo	Plan of Bâb Al-Futûh	105
	Bâb An-Nasr	108
	Burg Ez-Zefer. Plan of the East Gateway	109

	Burg Ez-Zefer. The East Gateway from Without	108
	Plan and Section of the Burg Muqattam	232
Old Cairo	Plan of the Fortress of Babylon in Egypt	44
Camber Castle	Plan of the Castle	233
Canterbury	The Castle. Plans, Section and Elevation of the Keep	73
	The Castle. Window of the Keep	112
	The West Gate of the City. Loophole	231
Château Gaillard	Plan of the Château	117
	The Donjon from the East	118
Chepstow Castle	Fitz Osbern's Hall from the South	81
	Marten's Tower	211
	Window in the Basement of Fitz Osbern's Hall	113
Chinon	Plan of the Tour du Coudray	125
Chipchase Castle	The Tower from the South-West	183
Colchester	Plan of the Balkerne Gate	39
	Plan of the Keep	71
	The Keep from the South-East	68
	Partition Wall in the Keep	63
	Window in the Keep	112
	Plan and Section of Fireplace in the Keep	115
	Fireplace in the Keep	107
Conisborough Castle	Plans, Section and Elevation of the Keep	96
	Window in the Keep	112
Constantinople	The Land Walls	41
Conway Castle	Plan of the Castle	156
	The Castle from the North-East	151
	The Castle from the South-West	151
Corfe Castle	The Castle from the South-East	67
	The Wall Walk through the Keep	107
	Plan and Section of the Middle Gateway	190
Craigmillar Castle	Plans and Section of the Tower	181
	The Castle from the South-East	176
Criccieth Castle	The Castle from the North-West	138
Deal Castle	Plan of the Castle	233
Denbigh Castle	Plan of the Gateway	195
	The Gatehouse from the North-East	200
Doune Castle	The Castle from the North-East	219
	The Tower House from within the Bailey	219
	Leaf of Iron Yett	201
Dover Castle	Plan of the Keep	80
Drawbridges	Method of Raising Drawbridge	203
Ellesmere Castle	Plan of the Castle	53

ILLUSTRATIONS ix

Elphinstone	The Tower from the South	219
Etampes	Plans, Section and Elevation of the Donjon	95
Exeter Castle	Plan and Section of the Gateway	104
Falaise	Loophole in Château	231
Flint Castle	Plan of the Castle	155
Foix	Plan and Section of the Château	174
	The Château from the West	175
Fréjus	Plan of the Porte des Gaules	34
Gisors	Plan of the Château	62
Hadrian's Wall	Plan of Borcovicium	37
Hall Windows		112
Harlech Castle	Plan of the Castle	160
	The Castle from the North	138
Haughley Castle	Plan of the Castle	53
Hittite City	From the Wall of the Great Hypostyle hall of the Ramesseum, Thebes	5
Houdan	Plans, Section and Elevation of the Donjon	91
	Window of the Donjon	112
Kenilworth Castle	Plan of the Inner Bailey	75
	The Keep from the South	78
	Plan and Section of the Battlements of the Keep	110
	Window in the Basement of the Keep	113
	Loophole in the Water Tower	111
	The Great Hall, Interior	228
Khorsabad	Plan of Gateway	10
Kirby Muxloe	Loophole in Gateway	231
Le Krak des Chevaliers	Plan of the Castle	136
Launceston Castle	Plan and Section of the Keep	58
	The Keep from the East	55
	Plan of the South Gateway	106
	The South Gateway from Without	78
Lewes Castle	Plan of the Castle	53
Leybourn Castle	Section and Elevation of the Gateway	194
Loches	Plan of the Donjon	69
	The Donjon from the South-West	69
	Window of the Donjon	112
Loopholes or Meurtrières		111, 231
London	The White Tower from the South-East	68
Longtown Castle	Plans of the Keep	97
	The Keep from the West	88
Lydford Castle	The Keep from the South	81
Maiden Castle	Plan of the Castle	26

ILLUSTRATIONS

Mallorca	Plan of the Castillo de Bellver	165
St. Mawes	Plan of the Castle	234
Medînet Habu	Plan of the Outer Gateways	6
	The Outer Gateways	7
	Crenellations on the Outer Wall	9
Manorbier	Loophole in Battlements of Curtain	111
Mycenæ	Plan of the Acropolis	2
Najac	Plan of the Inner Bailey	127
	The Château from the East	131
Newark, Selkirkshire	The Castle from the East	227
Nicæa	Plan of the City	42
Nîmes	Plan of the Porte Auguste	35
Old Cairo	Plan of the Fortress of Babylon in Egypt	44
Ortenberg	Plan and Section of the Castle	126
Parthenay	Porte St. Jacques	187
Pembroke Castle	Plan of the Castle	120
	Plan and Section of the Keep	121
	The Keep from the South	123
	The Keep, Interior, looking up towards the Dome	107
	Loopholes in the Battlements of the Keep	111
	Plan and Section of the Gateway	191
Pendennis Castle	Plans of the Castle	235
Pompei	Plan and Section of the Walls	23
	The Parapets	24
Provins	Plan and Section of the Tour de César	93
Queenborough	Plan of the Castle	166
Restormel Castle	Plan of the Keep	61
	The Keep from the West	56
	The Keep, Interior, looking towards the Gateway	56
Rhodes	The Ancient Walls	13
La Roche Guyon	Plan of the Château	119
Rochester Castle	The Keep from the North-West	77
	Fireplace in the Keep	107
Roumeli Hissar	The Castle from the West, looking across the Bosporus to Anadoli Hissar on the Asiatic Shore	82
	Plan and Section of the Black Tower	84
	The Black Tower from within the Castle	87
Sandgate Castle	Plan of the Castle	233
Sarzanello	Plan of the Castello	164
Senlis	Roman Walls	36
Sherborne	Plan of the Castle	102

ILLUSTRATIONS xi

Shutters	Single Shutter	205
	Double Shutter	205
Skenfrith Castle	Plan of the Castle	98
	The Keep from the South	88
	Loophole in Wall Tower	111
	Loophole at Base of Keep	113
South Coast Castles	Plans of Walmer, Deal, Sandgate and Camber	233
Tamworth Castle	The Castle from the South-East	55
	The Causeway	63
Tantallon Castle	Plan of the Castle	214
	The Castle from the South-East	212
	The Castle from the South-West	183
Tattershall	The Tower from the South-East	223
Threave Castle	The Castle from the South-East	184
	Plans and Section of the Tower	185
Tiffauges	Plan and Section of the Tour du Vidame	207
	Battlements of the Tour du Vidame	208
Timgad	Plan of the Byzantine Citadel	47
Tiryns	Plan of the Fortress	3
Totnes Castle	Plan of the Castle	57
Trematon Castle	Plan of the Castle	59
	Loophole in the Battlements of the Keep	111
	Loophole in the Gatehouse	111
Troy	The East Gate and a Portion of the Wall of the Sixth City	4
Troyenstein	Plan and Section of the Castle	215
Walmer Castle	Plan of the Castle	233
Warwick Castle	Plan of the Castle	177
	The Keep from within the Bailey	64
	Plans and Section of Cæsar's Tower	178
	Cæsar's Tower and the Gatehouse from within the Bailey	199
	Plan and Section of the Gatehouse and Barbican	197
	The Gatehouse and the Wall Walk on the Curtain	200
Visby	The Town Walls	172
Windows		112
Yett		201
York	The Multangular Tower from the South-West	8
	The Multangular Tower, Interior	8
	Plan and Section of Clifford's Tower	134
	Clifford's Tower from the North-East	132

PREFACE

MODERN researches point strongly to the conclusion that the art of fortification had reached a high state of development even at the dawn of history. Powerful military works, dating from the remotest periods, have been found in Asia Minor, in Greece, and in the basins of the Tigris, the Euphrates, and the Nile. And in the development of the art through the ages it is also clear that the peoples of these countries in their continuous conflicts with each other, and later with the legions of Rome, learnt and adopted the methods of attack and defence of their foes and in the course of time their military works became more and more similar.

With the rise of the Romans the development received a fresh impetus, and in the process of extending and consolidating their conquests the Romans built fortifications which, while being monuments of inventive genius and engineering skill, each adapted to its particular purpose, were also at any given period of great uniformity in fundamental principles of design.

For many centuries following the fall of the Roman Empire the nations of the West reverted to their previous state of barbarism and the development of the art was confined largely to the Empire of the East. But from about the eleventh century of our era the development again became widespread and general among the leading nations of Europe and the Levant. For it was the serious affair of every great leader to be familiar with the latest methods of attack and defence, since his safety and the safety of his followers depended upon his ability to forestall any surprise line of assault; and the Middle Ages afforded abundant opportunities for travel. The pilgrimages to Jerusalem, which took place from the fourth century of our era onwards, the Crusades, the pilgrimages to Rome, Compostella, Cologne, and other holy places, gave great scope for such investigations. Again private journeys, both at home and abroad, were made frequently by ecclesiastics and by laymen, and it is abundantly clear from their works that during these journeys they kept their faculties alert, and that on their return they made good use of their observations.

In the development of fortifications there were, naturally, variations of form in the several countries, consequent on differences of climate, available material, and national characteristics. Early advances were also made in one country and a tardy adherence to traditional practice is observable in another. But, none the less, the progress was, in the main, general. A survey and study of many of the castles of Europe and the Levant has led the writer to the conclusion that no clear grasp of the history of military architecture can be obtained from the study of examples in one country alone, and that more general treatment is essential to a proper understanding of the subject.

The object of this work is to trace the development of the art of fortification throughout Europe and the Levant generally from the period of the earliest historical examples down to the sixteenth century of our era; noting in their order, the salient features of the military works themselves as well as the siege operations employed against them. Since the design of the fortifications, the details of structure, and the methods of attack and defence employed were, in essence, the same whether related to a town, a castle, or even a camp, the developments are noted in order wherever they occur, and the fortifications described and illustrated are chosen from among those which retain their original features most complete.

There is no pretence here at anything approaching to exhaustive treatment of a subject so wide in a space so limited, and no one is more conscious of the deficiencies and omissions of the following pages than the author; the material must of necessity be selective. But it is hoped that the work may be a useful contribution to an important and fascinating field of study.

The Author wishes to express his thanks for valuable help to Sir Charles Leonard Woolley, M.A., who kindly read the first three chapters; to Dr. R. E. M. Wheeler, M.C., M.A., F.S.A., who read the fourth chapter; to Mr. A. W. Clapham, C.B.E., F.B.A., Sec.S.A.; and to Mr. R. Welldon Finn, M.A. He also gratefully acknowledges information obtained from some modern works on the subject. In particular the following: *L'Afrique Byzantine*, by Charles Diehl, Paris, 1896; *Manuel d'Archéologie Française*, by Camille Enlart, Paris, 1904; *Deuchen Burgen*, Berlin, 1889; and *Die Burgen Italiens*, Berlin, 1909, both by Bodo Ebhardt; *Military Architecture in England*, by A. Hamilton Thompson, M.A., F.S.A., London, 1912; the Inventories of the Royal Commissions on Historical Monuments, London and Edinburgh; *Arquitectura Civil Española*, by Lamperez and Romea, Madrid, 1922; *The Medieval Castle in Scotland*, by W. Mackay Mackenzie, London, 1927; and the monographs on Scottish Castles by Dr. W. Douglas Simpson, M.A. He desires especially to thank the numerous governing bodies and private owners, at home and abroad, who kindly granted him permission to examine the buildings in their charge. All the drawings have been prepared and the photographs taken by the author, and he has himself examined and surveyed most of the fortifications described.

SIDNEY TOY.

1 *Cloisters,*
    Temple,
        London, E.C. 4.
January, 1939.

CHAPTER I

THE ANCIENT FORTRESSES OF BABYLON, MESOPOTAMIA, ASSYRIA, ASIA MINOR AND GREECE

FROM the earliest times cities and palaces were surrounded by walls, often of enormous thickness and of great height; the walls being surmounted by walks with embattled parapets. Sometimes there were two or even three lines of such walls, and a palace or citadel within the innermost line; as in the cities of the Hittites in Asia Minor.

Some ancient fortifications, remains of which still stand or have been uncovered by modern excavation, date from periods so remote and attain such high degrees of perfection that, at present, it is not possible to assign the elements of the science to any definite age. The fortifications of Atchana, near Antioch, date from the nineteenth century B.C.; the first walls of Babylon probably from a still earlier period; and the curtain wall of Ashur, the ancient capital of Assyria, about 1600 B.C. The fortifications of Mycenæ and Tiryns in Greece and of what is known as the Sixth City of Troy in Asia Minor, all date from the Mycenæan Age, about 1500 to 1200 B.C.

When these defences stood upon a plain, as at Babylon, they were surrounded by a moat. When on a hill, as at Mycenæ and Tiryns, the approach was by way of a ramp, which was commanded by a tower at the head and exposed to close attack from the walls.

The gateways were often of great width and height and flanked by towers. Sometimes, as at Mycenæ and Troy, they were defended by a long approach, under attack from the walls on both sides. In the walls of Atchana there is a triple gateway; at Mycenæ and Tiryns double gateways were built; and later, as at Khorsabad, the gateways were of massive proportions and the passages through them were of great length and were intersected by cross chambers.

BABYLON

Of Babylon so much has been destroyed and the remains brought to light in modern times are so fragmentary that our knowledge of the city is largely derived from the descriptions given of it by Herodotus, Ctesias, and Strabo; and these descriptions apply more particularly to the City as remodelled by Nebuchadnezzar II, about 600 B.C. The more ancient wall was about 23 ft. thick, was strengthened at intervals by towers, and was surrounded by a moat.

CASTLES

MESOPOTAMIA

Their country being flat, the Mesopotamians always built on the plain, and their walls, consisting of a rampart 26 ft. high, built of mud brick, and surmounted by a wall, built of burnt brick and bitumen, were defended by a moat, canal, or river.

MYCENÆ

The acropolis at Mycenæ stands on a hill at the north end of the lower city. It is enclosed by a strong wall of massive hewn masonry, and is entered from the lower city by a ramp and a double gateway. The ramp is enclosed on either side by walls built of large blocks of stone and is defended by a strong tower at the top, on the right hand side of the gateway. The outer gateway, the "Gate of the Lions," is built of monolithic jambs, which incline inwards as they rise, and a lintel composed of a single block

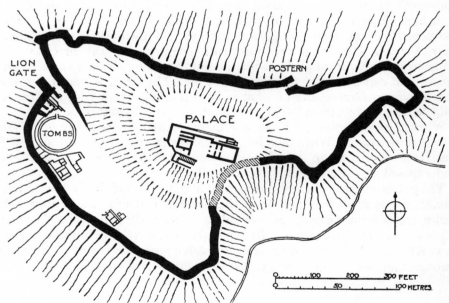

Mycenæ. Plan of the Acropolis. After Perrot and Chipiez.

of stone 16 ft. long and 3 ft. deep in the centre. Upon the lintel sculptured in high relief on a single block of stone 10 ft. high are the figures of two lions facing each other, one on either side of a column. Within the gateway there is a wide passage between two walls and at the other end of the passage there was another gate, now destroyed. On the north side of the acropolis, away from the lower city, there is a postern, also strongly defended, from

THE ANCIENT FORTRESSES 3

which a pathway leads down the hill to the open country. The ruins of the palace are on the summit of the hill in the middle of the acropolis.

TIRYNS

The hill on which Tiryns stands rises in two stages first to a long level platform and then up to another platform beyond and in line with the first. A wall of 'Cyclopean' masonry, 26 ft. thick and originally about 65 ft. high, is carried round the edge of the lower platform, up the slope on both sides, and round the edge of the upper platform; thus enclosing a long and relatively narrow space running north and south. A cross wall, built along the upper edge of the slope between the two platforms divides the fortifications into a citadel, on the highest point, and a lower ward or bailey. The palace and principal offices stand within the citadel.

Entrance to the fortress is by way of a ramp, or inclined path, which rises against the hill in the middle of the east side and leads to a gateway at the top. The ramp is commanded by a tower at the head and is so

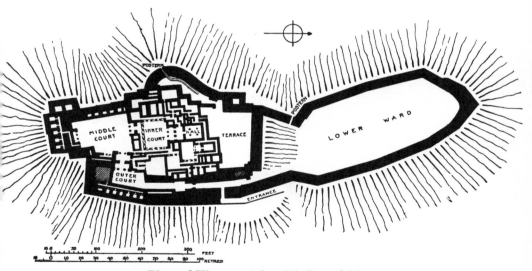

Plan of Tiryns. After W. Dörpfeld.

placed that the right or unprotected side of an advancing force is exposed to attack from the defenders on the walls above. The gateway, which had a gate at either end, opens on to the middle of a long passage running at right angles to it and having at one end the entrance to the lower ward and at the other the outer gate of the citadel. In order to reach the principal rooms of the palace, which are in the middle of the citadel, from this last gateway it is necessary to pass successively through another long passage, an open court, a second gateway, another court and a third gateway; each gateway being built at right angles to that preceding it. There are two posterns, both in the west wall; one at the foot of a long flight of steps down from the citadel, and the other near the south end of the lower ward.

TROY

At Troy the remains of nine cities, one buried beneath the other, have been brought to light, the earliest dating from about 3000 to 2560 B.C. The Sixth City, the walls of which were discovered by W. Dörpfeld in 1893–94, is of the Mycenæan Age and is probably the Troy of the Homeric epics.

The walls of this fortress are of great strength and, with a very steep batter on the outside face, rise even now in places to the height of about 20 ft. Along the faces of the wall, both inside and outside, there are vertical

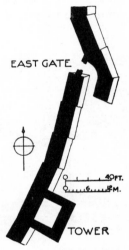

Troy. Portion of the Walls of the Sixth City.
After W. Dörpfeld.

offsets about 6 in. deep, running from the base to the summit at horizontal intervals of from 17 ft. to 27 ft. The panels of wall between the offsets are straight, and it is possible that this form of construction was adopted to avoid the curved surfaces which the contours of the curtain would otherwise require. Incidentally, the buttress-like appearance which the long vertical shadows of the offsets impart to the wall greatly increases its effect of strength. There are three towers in the remaining portion of this wall, one of which defended the south gateway.

The east gate is formed at a point where one portion of the wall slightly overlaps the other. The gate is placed on the inside face of the curtain and the approach to it passes between the two portions of wall at the overlap taking a right-angled turn in its course. The gate is therefore obscured from view from the outside (p. 4).

CITIES OF THE HITTITES

Of very early date also were the highly developed fortresses of the Hittites, that great people whose empire extended over the whole of Asia Minor, from the Ægean Sea to the borders of Mesopotamia. Ruins of these fortifications still exist but it is more particularly to the incised drawings on the walls of Egyptian temples that we owe our knowledge of their appearance when complete. A typical example is that of the City of Dapour, drawn on the south wall of the Great Hypostyle hall of the Ramesseum. The drawing itself dates from about 1280 B.C.

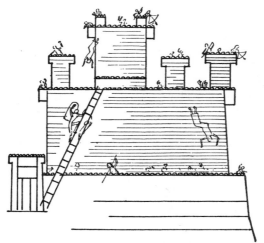

Hittite City. From the Ramesseum, Thebes.

The city is surrounded by two lines of curtain walls, and has a rectangular keep or fort within the second wall. There is a large gateway in the outer wall, flanked by a tower on either side. Turrets rise from the inner wall, and all the walls and towers are surmounted by embattled parapets. The city is being stormed by the Egyptians, and the Hittites are fighting from the battlements of all three lines of fortification. Hoards, temporary wooden platforms, are projected out from the parapets of the inner wall and from those of the keep, and men are fighting from them. The weapons used are bows and arrows and lances. The Egyptians advance under the cover of long shields, flat at the upper end and rounded at the other, and the attack is by means of scaling ladders. Some of the besiegers have evidently reached the battlements, for they are being hurled down from them by the defenders.

Substantial remains of Hittite fortified cities, defended by two or three lines of strong walls and having a citadel within the innermost line, still exist in various parts of Asia Minor, as at Sinjerli, Hamath, and Carchemish. Sinjerli was circular in plan and was surrounded by a double line of walls.

At Carchemish, on the Euphrates about seventy miles north-east of Aleppo, there were three lines of defences in succession, the Outer Town, the Inner Town, and the Citadel; the citadel being backed against the river. The walls of the Inner Town, where they have been excavated on the east, are 5.80 m., or 19 feet thick, and are built with external and internal vertical offsets, much in the same manner as those of the Sixth City of Troy. The gateways were flanked by towers and the passages through them intersected by large chambers and halls.[1]

EGYPT

At the southern extremity of Egypt, near Wadi Halfa, are the remains of three border fortresses, said to date from about 2000 B.C.; they are all of purely military character. One of them, at the island of Uronarti on the Nile, was built of very thick walls supported at frequent intervals by massive buttresses, constructed of brick with timber bonding.

The palatial buildings in Egypt, known as temples, such as the group at Karnak, the Ramesseum, and Medînet Habu, all at Thebes in Upper Egypt, are surrounded by fortifications, often of great strength.

The Ramesseum, built about 1280 B.C., consists of a long rectangular suite

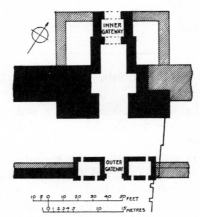

Medînet Habu.   Plan of the Outer Gateways.   p. 9.

[1] Vide "Carchemish. Part II. The Town Defences," by C. L. Woolley. London, 1921.

Medînet Habu, Thebes.
The outer Pylons, or Gateways, from without.   p. 9.

Multangular Tower, York. Exterior from the South-West. p. 40.

Multangular Tower, York. Interior, showing remains of the cross wall in centre. p. 40.

Medînet Habu. Crenellations of the Outer Wall.

of courts and halls, arranged one behind the other, entered by a massive pylon, or gateway, and surrounded by a girdle wall.

Medînet Habu was built about 1200 B.C., partly across the site of a smaller structure of about 1500 B.C. The new work consisted of a rectangular temple or palace of similar design to the Ramesseum, and a large fortified area surrounded by a double line of walls within which the palace stood. The entrance was by a fortified double gateway, still in good state of preservation (pp. 6, 7). The outer gateway is relatively small; it is defended by a guard-room on either side. The inner gateway known as the Pavilion of Rameses III, is a powerful structure, three storeys in height and flanked by strong towers. Above the gateway are two tiers of chambers, and there are other chambers in the towers with openings commanding the long passage to the gate.

Portions of the outer wall still remain to the full height, including the original crenellations and crenellations remain on the summit of the inner gateway. These parapets have semi-circular shaped merlons, similar to those represented on the walls and towers of the Hittite city (p. 5), and on drawings of the cities of Assyria.

ASSYRIA

The fortifications of Assyria differ very little in character from those of the Hittites. On the stone slabs from the palace of Ashur-Nasir-Pal at Nimrûd, now in the British Museum, are representations of cities besieged by the king about 880 B.C. These cities are defended by powerful walls, which have walks with embattled parapets on their summits, and are strengthened at frequent intervals by towers; often there are two or even three lines of walls. The gateways are wide and high, have usually round but sometimes flat heads, and are flanked by a tower on either side.

The city of Khorsabad, near Mosul, of which there are extensive remains, was built 722–705 B.C. It is of rectangular plan and was defended on all sides by a wall 79 feet thick, with wall towers at frequent intervals. The palace with its halls, courts and sacred precinct, stands on the north side of the city and partly projects beyond the line of the north wall, which is carried out around it. There was one gateway in the north wall of the city and two in each of the other walls. The gateways were all alike and were formidable structures, each flanked by a tower on either side. They extended inwards 60 feet beyond the inside face of the wall and the passage through, intercepted at two points by large cross chambers, was about 155 feet long. In front of each gateway there was a barbican, 150 feet wide by 82 feet long, with an outer gateway.

The city of Madaktu, as shown on a stone slab of about 650 B.C., now in the

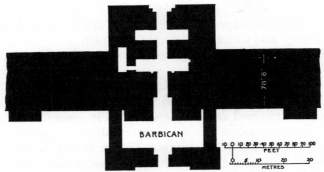

Khorsabad. Plan of one of the city gates.
After Perrot and Chipiez.

British Museum, was surrounded by a single wall with wall towers at frequent intervals and gateways, flanked by towers, at convenient points. It is also defended on the one side by a wide river, which curves half-way round it, and on the other by a small stream or moat. There is a wide space between the walls of the city and the river, and on this ground are houses, interspersed with trees, and two castles. Both the castles stand close to the river, one of them rising directly from the ground and the other, a much larger structure, standing on a mound. Within the city are streets of houses, each house having one or two towers, and trees are planted in all parts.

About 600 B.C. Nebuchadnezzar strengthened the fortifications of Babylon by increasing the thickness of the existing wall on the outside, digging a new moat, and constructing an entirely new line of defences, consisting of a wall and a moat, inside the old walls. The city was now protected by a double line of walls and moats, the outer moat having a lining of burnt bricks, set in asphalt. The full thickness of the outer wall, as measured from the ruins themselves, is 85 feet 8 inches. This dimension agrees closely, if not exactly, with the measurement given by Herodotus.[1] The figures that author gives for the height and extent of the walls must be taken simply as expressive of great proportions; exact measurements on such a scale would involve a complicated survey unreasonable to expect.

The Ishtar gate at Babylon, crossing the sacred way near the palace, was also built about 600 B.C. and still stands to a considerable height. Its walls are decorated in low relief with large figures of bulls and dragons, executed in glazed brickwork. Here the long passage through the gateway is intersected at three points by large cross chambers.[2]

Nebuchadnezzar built a new wall around the Temenos at Ur of the Chaldees. This wall was composed of two solid portions, each 3.25m. thick, built 5.20m. apart and connected at intervals by strong cross ribs; the total thickness being 11.70m. or 38 ft. 4 in. It is strengthened on both faces by shallow buttresses, spaced closely together.[3] The four principal gateways are set back from the general face of the wall in such a manner as to form wide spaces in front, exposed to attack from the walls on three sides. There are chambers on either side of the passage through the gateway.

ATHENS

At Athens the acropolis is built on the summit of a hill which stands within and is completely surrounded by the lower city. About 480 B.C. a wall was built on the top of the precipitous hill round the acropolis, with a gateway at the west, the only accessible side. At the same period another wall, with gateways on the ancient roads, was carried all round the lower city. The gateway at the principal approach to the lower city, called the Thriasian gate, was rebuilt about 350 B.C. and renamed the Dipylon, or double gate. The Dipylon was built on the principal of that at the entrance to the acropolis at Mycenæ, having outer and inner gates with an open passage or court between them; here, however, there were twin gates at either end of the court (p. 12). Both pairs of gates were flanked by towers and the court was enclosed by thick walls on either side. An enemy who had entered the outer and was checked by the inner gates would be under attack from all sides of the court.

[1] *Herodotus: Clio* 178.
[2] *Vide* W. L. King. *A History of Babylon.*
[3] "The Excavations at Ur," by C. Leonard Woolley. *The Antiquaries Journal*, Vol. V. 1925, p. 360.

It was largely due to the strength of the Dipylon that Athens was able to resist the attack of Philip V, King of Macedonia, in 200 B.C. Finding that his design to take the city by surprise had failed, Philip decided on open attack and advanced towards the Dipylon. An Athenian force went out to meet him and great numbers of citizens stood upon the walls of the city to watch the conflict. Philip, observing this crowd of spectators and desirous of making a parade of his prowess, advanced with a small body of horse, forced back the Athenians and rushed the outer gates. The fight was now confined within the limits of the court between the two barriers, and the king owed his safe retreat to the fact that the defenders on the towers and flanking walls feared to fire on the mêlée below lest they should strike their own men.

Henceforth the citizens kept within the inside barrier of the Dipylon, and Philip, not wishing to be entrapped again, drew off his army.[1]

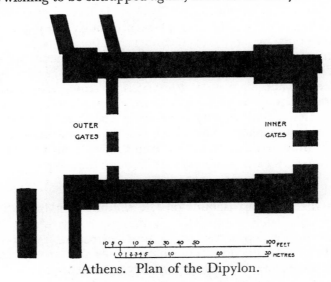

Athens. Plan of the Dipylon.

RHODES

A development in military construction occurred at Rhodes, in the building of the ancient curtain wall of the city about 400 B.C. On the side towards the city this wall formed an arcade of tall arches, vaulted just below the wall walk (p. 13). The maximum thickness of the wall was 15 ft. The arches were 15 ft. wide by 10 ft. 6 in, deep, and the piers were 15 ft. wide.[2] This scientific method of construction was followed later by the Romans and the Byzantines. Its advantages were threefold; a great saving of material; in

[1] Livy. *History of Rome*, XXXI, 24.
[2] Philon de Byzance (fortifications) par A. de Rochas-Aiglum et C. Graux. *Revue de Philologie*. Paris, 1879.

the event of a breach being made the damage was localised and its repair facilitated; and the arches themselves were useful for the accommodation of guards.

The strength of the fortifications of Rhodes and the military skill of its defenders were put to severe test in the memorable siege of the city by Demetrius Poliorcetes in 305 B.C.; and the victory was with Rhodes. Demetrius had brought against the city the most powerful siege engines of his time, including petrariæ—heavy stone-throwing engines—and battering-rams which were each 150 ft. long and worked by a team of 1,000 men. He had assailed the city from the harbour by floating batteries, and from land by his

Rhodes. Walls of the Ancient City.

renowned "Helepolis" or huge siege tower. Yet, despite his skill and his resources the king was compelled to raise the siege after it had been carried on relentlessly for a whole year.

WEAPONS, SIEGE ENGINES AND METHODS OF ATTACK AND DEFENCE

The weapons in use about 1280 B.C., as shown on the stone slabs of the Ramesseum and elsewhere in Egypt, are bows and arrows and lances.

Scaling ladders were used to carry the walls of fortresses. By 880 B.C. the art of attack and defence had made great progress. The Assyrian bas-reliefs of that period show men rushing to the attack of a fortress with a lance in one hand and a shield, either round or rectangular, in the other. Archers advance under cover of mantlets, held in front by attendants; sometimes one attendant grasping the mantlet with one hand and holding a sword in the other, to two archers. Mantlets were light screens constructed on a wooden frame and of sufficient height and width to cover the party.

The defenders of the fortress fought from the battlements of the walls and towers, and from hoards projected out from the battlements, shooting and hurling a rain of arrows, stones and firebrands at the assailants and their engines.

The walls are attacked either by battering rams or by men sapping at the base with crow-bars. The rams, shod with iron and often in pairs, were set up in a strong timber framework, which moved on wheels and was entirely covered by raw hides as a protection against fire. The whole machine was mobile and those working it were under cover. In order to check the ram the besieged endeavour to grapple its head in the loop of a chain they lower down from the wall; and the besiegers, on their side, seize the chain with hooks and tug at it with all their might.

There were also siege towers, which were built of timber, covered with hides, and mounted on wheels; they were of two or more storeys and from the top storey men fought on a level with those on the battlements of the fortress. According to Vitruvius, Diades, who flourished under Philip, King of Macedonia, 360–336 B.C. claimed to have invented ambulatory towers which could be dismantled and taken about from place to place in pieces.[1]

Fire played an important role in an assault, and flaming torches were thrown both by besiegers and besieged. Torches hurled at a fortress and water thrown down from the walls to quench them are shown on the reliefs of 880 B.C. But during the following century the use of this powerful weapon, especially by the besieged, appears to have grown enormously. In the assault on Lakish by Tiglath-Pilesar III, c. 750 B.C., immense numbers of flaming torches were thrown from the walls of the city on the advancing troops and their siege engines. Men with long-handled ladles, working under the protection of small towers on the top of the engines, fought against the fire by pouring water over the affected parts. In this attack hand-slings were used by both sides. Intimidation by resort to "frightfulness" was pursued remorselessly, some of the prisoners being tortured or flayed alive, and others impaled within sight of their friends in the city.

The principles of attack and repulse by means of mines and counter mines dug beneath the walls of fortresses had reached a high state of development by the end of the sixth century B.C. At the siege of Barca in Libya, c. 510

[1] *Vitruvius de Architectura*, XIX.

B.C., the Persians excavated underground tunnels that reached to the walls. Among the Barcæans there was a skilled worker in brass who took a brazen shield and, carrying it round within the wall, applied it here and there at places where he thought the workings might be. Where there were no works the shield made no sound, but at places where excavations had been made the shields sounded. By countermining the Barcæans broke into the enemy's works and slew the miners there.[1]

Mines, excavated by besiegers, usually commenced some distance from the walls and had one of two objectives; to effect direct entry by means of tunnels which passed beneath the wall and opened out within the fortress; or to make a breach in the wall itself. The second, perhaps a later phase, is referred to by Polybius. The procedure was to excavate a large cavity in the base of the wall, propping and strutting with timbers as the work proceeded. When the cavity was large enough the timber was fired, the men withdrew, and, if the work had been well done, on the consumption of the wood the wall above collapsed.

By the end of the fourth century B.C. petrariæ had been introduced and great strides had been made in the design and development of battering rams and siege towers.

---

[1] *Herodotus Melpomene*, 200.

CHAPTER II

FORTIFICATIONS OF GREECE AND ROME: 300 B.C. TO 200 B.C.

DURING the third centry B.C. the art of war made rapid progress, due in no small measure to the inventive genius of that brilliant mathematician and engineer Archimedes. Born in 287 B.C., Archimedes lived at Syracuse and designed many powerful engines of war for its ruler Hiero II. He also effected many improvements in the fortifications of the town. So successful were the measures of defence adopted under his directions that the combined assaults of the Roman army and navy on the town in 215 B.C. were repulsed again and again, until the forces drew off and substituted blockade for attack. "Such a great and marvellous thing does the genius of one man show itself to be when properly applied to certain matters. The Romans had every hope of capturing the town if one old man of Syracuse were removed."[1]

It is quite clear, however, from the tactics adopted and the engines used by the Romans, by Hannibal, and by Philip V of Macedonia, that the progress in the art during this century was general. In their attack of Lilybæum in Sicily in 250 B.C. the Romans had all their siege engines destroyed by fire hurled at them from the battlements. They then determined to accomplish by famine what they had failed to do by assault and dug a ditch and put up a stockade all round the city. They also built a wall round their own camp,[2] a precaution which, if not common at the time, hereafter became the general practice. At the siege of Capua in 212 B.C. the Romans, advancing from three different points, completely surrounded the city by their works, which consisted of a rampart and two ditches, with strong forts at intervals all along the line.[3] Some towns themselves were protected with two or more lines of moats; at Sirynx in Asia in 210 B.C. there were three wide and deep moats, outside the wall, each defended by a stockade.[4]

The Romans possessed a strong navy, including ships equipped with grabs and other mechanical devices. In the fleet taking part in the siege of Syracuse in 215 B.C. were vessels, called sambucæ, which could be brought up to the sea walls and, by means of ladders raised up by ropes and pulleys, could land their men directly on to the battlements of the town. To counter the attack from these and other vessels of the Roman fleet, Archimedes

[1] *Polybius*, Book VIII, 7. [2] Ibid. Book I, 42.
[3] *Livy*, XXV, 22. [4] *Polybius*, X, 31.

ordered the construction of a number of stone-throwing engines of various sizes, the larger for long distance ranges and the smaller for short ranges. One night, under cover of darkness, the Romans brought their ships so close up to the walls as to be too near for even the shortest range engine. But the "old man of Syracuse" was not defeated. He pierced the curtain wall itself with loopholes, 6 ft. high by 4 in. wide and spaced at short intervals apart. Through these loopholes the archers, themselves secure behind the wall, so disabled the enemy that he retired.[1] It was more by stratagem than by force of arms that three years later Syracuse fell, and its defender Archimedes fell with it.

Sometimes when one part of the curtain wall, under severe attack from battering rams and mines, was in danger of collapse the defenders built up an entirely new wall within the city across the damaged part, as was done during the siege of Abydus in 201 B.C.[2]

Gateways, always vulnerable points in the defences of a fortress, received particular attention. They were closed by gates and later also by portcullises. It is not possible to give any definite date to the introduction of the portcullis, but it is referred to in a treatise on military tactics of the fourth century B.C. attributed to Aeneas Tacticus. This author writes: "If a large number of the enemy come in after these and you wish to catch them you should have ready above the centre of the gateway a gate of the stoutest possible timber overlaid with iron. Then when you wish to cut off the enemy as they rush in you should let this drop down and the gate itself will not only as it falls destroy some of them, but will also keep the foe from entering, while at the same time the forces on the wall are shooting at the enemy at the gate."[3]

The portcullis comes into prominence at the town of Salapia, near Barletta, during the Second Punic War, in 208 B.C.

Hannibal having got possession of the signet ring of Marcellus, who had been surprised and slain in ambush, attempted to use it to obtain entry into Salapia unopposed. In the name of Marcellus and signed with his seal Hannibal wrote a letter to the citizens informing them that they were to expect him on the following night and that the soldiers in the garrison were to hold themselves in readiness for any service he might require of them. But the Salapians had been forewarned that Marcellus was dead and that no credence was to be given to letters written in his name. Preparing therefore for an enemy they dispersed their men along the wall and placed a particularly strong guard at the gate at which they expected Hannibal to appear.

The Carthaginian vanguard consisted of Roman deserters, who, in the Latin tongue, ordered the guard to open the gate, as the consul had arrived. With a great pretence of excitement and haste the guards proceeded to obey.

[1] *Polybius*, VIII 5.
[2] Ibid. XVI, 30.
[3] *Aeneas Tacticus C.*, XXXIX. Ed. Capps, Page and Rouse, 1933.

The gate (*cataracta*) worked up and down like a sluice gate and was operated by levers and ropes. It was indeed what was later called a portcullis. Eagerly the enemy pressed through the gate, but when about six hundred of them had entered, the guards released the ropes and the gate fell with a crash, blocking all further entry. Thereupon, while the defenders on the walls and towers rained stones, spikes and javelins on the enemies without, the citizens fell upon those who had entered and who, carrying their arms suspended from their shoulders, were taken completely by surprise. And Hannibal was forced to retire.[1]

SIEGE ENGINES AND SIEGE OPERATIONS

Great advance had also been made in the manner of conducting an assault on a fortress and in the design and construction of siege engines. In the attack on Syracuse in 215 B.C. Archimedes constructed powerful petrariæ which were the prototypes of disappearing guns. These engines when not in action lay concealed behind the battlements. But when brought into play a great beam suddenly reared up into view, swung round on its axis and cast a stone weighing as much as 5 cwts. Balls of lead were also thrown by these engines.[2]

Ballistæ, engines for throwing stones up to about 56 lbs. in weight, and catapults, engines for throwing arrows and firebrands, were also used at this period, as were mantlets, often extended into long screens. The siege towers were sometimes of many storeys.

In the attack on Echinus by Philip of Macedonia in 211 B.C., the king, having selected as his point of attack one section of the city wall with a tower at either end, proceeded to erect elaborate works against that section. He built two tall siege towers with battering rams, one opposite each of the wall towers, and between his siege towers, connecting one with the other, he constructed a rampart of earth and a screen or stockade. In the gallery behind the stockade he set his miners to work driving two tunnels towards the wall, while other miners were employed in digging a system of underground passages between the camp and this gallery, so that his men could pass to and fro safely. The siege towers were each of three storeys, and in addition to fighting men, contained catapults, fire extinguishing equipment, and, in the ground storey, men employed in clearing the way in front of the towers so that they could be moved forward. In addition to these works Philip had three batteries of ballistæ trained on to the city, which finally surrendered to him.[3]

Many cities were carried simply by escalade. In their assault on Illiturgi in Spain in 207 B.C. the Romans observed that the highest part of the city stood

[1] *Polybius*, X, 33. *Livy*, XXVII, 28.
[2] *Polybius*, VIII, 5.
[3] Ibid, IX, 41–42.

on a precipitous cliff but was otherwise unfortified. By the use of spikes, which they drove into the cliff above them as they mounted, those behind supporting those above and those above pulling up their companions below, they swarmed up the cliff and took the city. The walls at New Carthage were carried by escalade in 210 B.C. As a defence against escalade specially made prongs were used to prevent the fixing of the ladders, and iron grapples were let down from the battlements, seizing the assailants and lifting them up the face of the wall.[1]

Grappling cranes were constructed by Archimedes during the Siege of Syracuse. These cranes stood on the sea walls and when a Roman ship approached let down an iron hand, which clutched the ship by the prow and raised it until it stood upright on its stern; they then suddenly released it, and the vessel fell back into the water to be either overturned or swamped. Men also were clutched from the deck of a ship and dropped into the sea by this engine.[2]

Difficult problems of transport were often encountered and solved. The genius of Hannibal in this respect was amply demonstrated during the siege of Taranto in 212 B.C. Taranto is built on a peninsula which stands across the mouth of its great natural harbour, called the Mare Piccolo, leaving only a narrow channel between the point of the peninsula and the land on the other side. The citadel was built at the point, commanding the channel, which at the time of the Punic wars was the only entry into the harbour. Hannibal had obtained possession of the lower city and had thrown up works between it and the citadel. But, finding the latter too strong to be taken by storm, he decided on a blockade. While, however, supplies were reaching the citadel by sea a blockade could not be effected. His own fleet was at Sicily and the ships belonging to Taranto were bottled up in the harbour, held in by those in command of the channel. Determined to get these ships out Hannibal drew them from the water by means of engines, loaded them on to waggons, and had them transported across the neck of land and launched in the sea on the other side. They then sailed round opposite the citadel and completed the blockade.[3]

"Frightfulness" as practiced by the Romans appeared to be actuated as much by ruthless revenge as by the desire to intimidate. Polybius says that following the taking of the lower city of New Carthage "when Scipio thought that a sufficient number of troops had entered he sent most of them, as is the Roman custom, against the inhabitants of the city with orders to kill all they encountered, sparing none, and not to start pillaging until the signal was given. They do this I think to inspire terror, so that when towns are taken by the Romans one may often see not only the corpses of human beings, but

[1] *Livy*, XXVIII, 19.
[2] *Polybius*, VIII, 6, 7.
[3] *Livy*, XXV, 11.

dogs cut in half and the dismembered bodies of other animals, and on this occasion such scenes were very many owing to the numbers of these in the place."[1] On the capture of Illiturgi by the Romans Livy says: "They butchered all indescriminately, armed and unarmed, male and female. Their cruel resentment extended even to the slaughter of infants."[2]

---

[1] *Polybius*, X, 15.   [2] *Livy*, XXVIII, 20.

CHAPTER III

FORTIFICATIONS OF THE LEVANT, ROME, AND WESTERN EUROPE
200 B.C. TO 30 B.C.

MILITARY architecture had now become a special science having schools at important centres, as at Rhodes; and during the two centuries preceding our era many treatises were written on the subject. Philo of Byzantium, who flourished about 120 B.C., wrote a treatise on mechanics and military architecture, which, judging from the fragments still extant, must have been a most comprehensive work.[1] From this treatise we learn the principles of the art as taught and practiced in the second century B.C.

The plan adopted for any fortress, Philo says, must be decided upon only after a careful inspection of the site it is to occupy, as the salients, inclination, and curves of the curtain walls are determined by the nature of the ground on which they are to stand. He describes several plans as the "ancient," the "saw-shaped," the plan with concaved walls, and the plan with double walls. Curtain walls should be at least 15 ft. thick, built in gypsum and well bonded; to prevent escalade they should be at least 30 ft. high. Sometimes the side of a fortification most exposed to attack is protected by two walls, spaced from 12 ft. to 18 ft. apart, and joined at the top by a vault or timber roof. Some curtain walls are embattled, but have no wall walks. In time of siege temporary platforms of timber may be put up behind these battlements, and removed when necessary. Even if the enemy is able to scale these walls, he will find on arrival at the top that he can proceed no further, and, being himself an easy target, he is faced with either death or retreat.

The curtain wall should be 90 ft. from the houses of the town, thus providing a road for the easy transport of engines, vehicles, etc., and the passage of reinforcements along the whole line of defence, and, also, in case of need, a space for digging an internal intrenchment.

Towers must be of a form suitable to the position they occupy in the wall. If angular, they should be so set as to present a projecting angle in the centre, so that blows may be received on a salient angle. When round, the face stones should be cut from wood templates to expedite construction. Towers should not be bonded to curtain walls; or, owing to the inequality of weight, fissures will develop and endanger their stability; they should be protected by a base work, or bastion, to prevent the approach of sappers. Both walls

[1] Philon de Byzance. *Vide* Note p. 12.

and towers should stand upon solid foundations and their upper courses should be set in gypsum and strengthened by iron clamps, run with lead. In positions most exposed to attack from siege engines the walls should be faced with hard stones or stones with salient bosses and well tailed into the body of the wall. Timber ties of oak should be buried in the walls of both curtains and towers, the timbers being placed end to end and forming horizontal chains at vertical intervals of 6 ft. The presence of these ties greatly facilitates the repair of any part of the wall which may be damaged.

Posterns are often built on the sides to facilitate the making of sorties. They are so arranged that the soldiers when in retreat are not obliged to turn to the left, thus exposing their unprotected right sides; but one file making a sortie from postern A will re-enter by postern B, and all the other files will adopt the same course. Some posterns will be constructed obliquely through the wall, while others have an elbow turning; but all should be so designed as to be out of the range of stone-throwing engines and obscured from view from the outside.

Great attention should be paid to the outworks. There should be an advanced wall and at least three lines of ditches; the spaces between the ditches being protected by palisades and planted with thorns. In front of the advanced walls empty earthenware jars should be buried. These are placed in an upright position with their mouths upward, stopped with seaweed or imperishable grass and covered with earth. Troops may then pass over the jars with impunity, but the engines and timber towers brought up by the enemy will sink into them.

Vitruvius, writing about 30 B.C., says that the plan of a stronghold should not be square, or have sharp angles, but should be polygonal, so that the movements of the enemy might not be obscured by salient corners. The fortification should be surrounded by uneven ground to make approach difficult; and the roads leading to the gates should be winding and turn at the gate so as to expose the right side of an attacking force. Foundations should be of greater thickness than the walls they support, and the walls should be of sufficient width to permit of two armed men passing each other freely on the wall walk.

Towers should be spaced not more than an arrow's flight apart in order that the wall between them may be swept from end to end by the engines on either tower. The wall walk from one section of the wall to the other must be carried across the inside of the tower by a timber bridge only. Then, if one section of the wall is carried by storm, that section can be isolated by removing the bridge in the tower at either end. Vitruvius agrees with Philo that walls should be consolidated with timber bonding, and recommends scorched olive wood; a material which, he says, is imperishable and will remain unimpaired either when buried in earth or immersed in water.[1]

[1] *Vitruvius: de Architectura.* V.

FORTIFICATIONS OF THE LEVANT, ROME, AND WESTERN EUROPE 23

Among the most interesting military works of this period are the ruins of Pompei. The fortifications of Pompei have the rare historic value of immunity from alteration since A.D. 79, when the city was overwhelmed. It is obvious that all the defences are of a period anterior to that catastrophe.

The Herculaneum Gate at Pompei dates from about 100 B.C. It has a wide central carriage way flanked on either side by a footway and is about 60 ft. long. The carriage way was closed by a portcullis on the outer side of the gate, and by wood doors on the city side. The footways were closed by

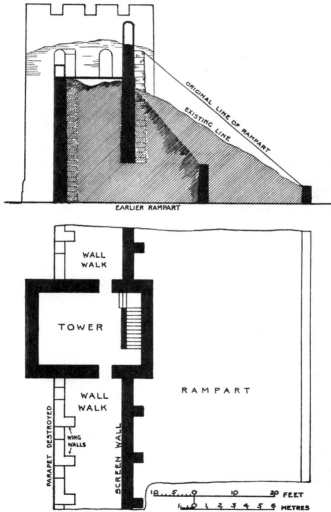

Pompei. Plan and Section of the Walls. After A. Maiuri. p. 24.

iron doors on the outer and by wood doors on the inner side. The middle portion of the gateway formed a large hall or court which stretched across all three passage ways.

The city wall is of various periods, from about 400 to 100 B.C.[1] As finished at the latter period it was composed of outer and inner facings of

Pompei. The Parapets.

stone and a core of earth; it was 20 ft. thick and rose to the height of about 32 ft. to the wall walk (p. 23). On the inner side the wall was supported by buttresses, placed about 10 ft. apart, and was backed by an inclined rampart of earth. The outer parapet has disappeared, but from the arrangement at its base of short buttresses projecting into the core of the wall at regular intervals it appears to have been provided with wing walls. Wing walls projected inwards from the left ends of the merlons of battlements, and, according to Procopius, were for the protection of the left flank of the men firing through the embrasures, in the event of another part of the wall being carried by assault. Traces of such walls have been found at Pevensey Castle and elsewhere, and A.D. 536 they were introduced into the fortifications of Rome.[2] In lieu of an inner parapet there is a screen wall which was carried up to the height of about 14 ft. above the wall walk and protected the city from the missiles fired from siege engines. Rectangular towers, extending through the full thickness of the wall and projecting out on either side, are built at intervals along the line of this fortification. They are open to the wall walk by doorways on both sides.

WESTERN EUROPE

The fortifications of the southern parts of Western Europe at this period had, in certain places, already reached a state of development as far advanced as those of Rome and the Levant. At Tarragona, in Spain, there still exist gateways, towers, and portions of massive and lofty walls which were built

[1] *Vide Monumenti Antichi R. Accademia Nazionale dei Lincei*, Vol. XXXIII. 1929.
[2] *Procopius: History of the Wars*, Book V, C. XIV, 15.

about 210 B.C. on still earlier foundations and were added to in the time of Augustus. The oldest portion of these powerful fortifications, dating from prehistoric times, is constructed of huge blocks of stone, measuring about 12 ft. long by 5 ft. high.

From Cæsar's account of his assault on Marseilles it is clear that the defences of that independent city were also powerful stone structures.

Further north the fortifications, though more crude, were still of great strength. In his campaigns in Gaul 58 B.C.–49 B.C. Cæsar found that the Gallic cities were surrounded by walls, built of large stones and earth and bonded by an ingenious system of timber ties. The outer ties were exposed on the face of the wall, alternating with the courses of mansonry. These walls, while not unpleasing in appearance, offered enormous resistance to the assaults of his battering rams.[1]

In Britain, before the advent of the Romans, the inhabitants lived in towns, built on hill-tops or on the level, and defended by ramparts and ditches. Sometimes the sites chosen were defended by nature on two or three sides, as on a promontory. In that case it might be necessary only to throw up artificial works across the neck of the headland. Otherwise the ramparts and ditches were carried all round, as at Maiden Castle, Dorset, one of the finest fortifications of its kind in existence.

Scattered throughout Scotland and now in ruinous condition are a large number of forts called brochs. Such of these structures as have been scientifically examined appear to date from the second century of our era, and others may be of much later date. But brochs are of Celtic origin and were probably being built long before the Roman occupation of Britain.

Brochs are tower-like structures, built of dry stone rubble. The walls are from 12 ft. to 15 ft. thick at the base, generally have a deep batter on the outside, and were originally from 45 ft. to 60 ft. high, enclosing an open courtyard. The entrance, the only external opening, is low and narrow, and the passage from it, frequently defended from a mural guard-room on one side and from apertures in the flagstone roof above, leads straight through to the courtyard. Chambers and stairways are formed in the thickness of the walls; the stairways as they rise spirally round the walls giving access to tiers of galleries. The galleries are built one immediately above the other and are lighted from the courtyard. In the courtyard there is often a hearth and occasionally a well and an underground cellar for stores. The best preserved example is that at Mousa in Shetland. This broch is circular, has a deep batter on the outside and, although now incomplete at the top, is still 45 ft. high.[2]

Walls composed of stone bonded by timber ties, like those of the native Gallic cities, are not often seen in British works, though there can be little

[1] *Cæsar: De Bello Gallico*, VII, 23.
[2] *Vide The . . . Scottish Brochs* by A. O. Curle. *Antiquity*, Vol. I, 1927.

doubt but that stone and timber were used where both could be procured. In stone-built forts, as at Grimspound and Cow Castle, the walls are built without mortar and consist of a rubble core faced with large stones. In districts where timber was plentiful and stone scarce, large stakes densely intertwined with branches of trees and thorns were sometimes the materials used in the defences.

The strength of the gateways consisted largely in the difficulty of access. The entrance was placed near a precipice, as at Mount Caburn in Sussex, or was masked by cross ramparts involving sinuous passages of approach, as at Maiden Castle and Hod Hill; or again the passage of approach was protected by an outwork on either side, as at Blackbury Castle, Devonshire. Sometimes the principal enclosure or citadel was protected by one or two outworks, as at Winkelbury, Wilts. and Bury Castle, Somerset.

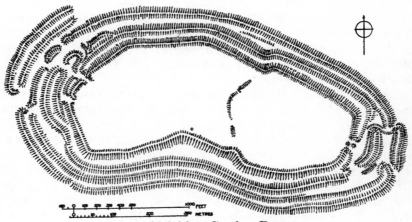

Plan of Maiden Castle, Dorset.

Maiden Castle stands on a site which was occupied as early as 2000 B.C. The eastern third of the site was fortified about 300 B.C., and the fortifications were extended westward to their present dimensions about 100 B.C. The castle covers an extensive oval area on the crown of a hill, 432 ft. above sea level, and is defended all round by several lines of ramparts and ditches Recent excavations have shown that the ramparts consist of earth, chalk, clay, and rubble, faced on the inside with stone and the facing itself buttressed at intervals by upright timber posts. On the summit are a series of post holes, spaced at from 4 ft. to 2 ft. 6 in. apart. These holes belong to a defence of a later period; the original wall, of which a portion still remains, was of stone.[1]

[1] *Vide* "The Excavation of Maiden Castle, Dorset," by Dr. R. E. M. Wheeler. *The Antiquaries Journal*, Vol. XVI, 1936.

FORTIFICATIONS OF THE LEVANT, ROME, AND WESTERN EUROPE

SIEGE ENGINES AND SIEGE OPERATIONS

The engines of attack and defence were considerably improved during this period. Philo speaks of having seen an engine, invented by Ctesibius, which threw stones by means of compressed air; and Vitruvius, in the course of his treatise on the construction of siege engines, describes a ram which, rotating in a channel by means of pulleys, acted as a boring machine. During the siege of Marseilles in 49 B.C. the projectile engines in the city were so powerful that they threw iron pointed poles 12 ft. long with such force that the poles, having pierced through four rows of mantlets, were brought to rest only by burying themselves in the ground.[1] Sickle-shaped hooks, fixed at the end of long poles, were used at the siege of Ambracia in 190 B.C. for pulling down the battlements of the city.[2]

The "tortoise" was used by the Romans at the capture of Heracleium in 169 B.C. Three picked maniples were employed. The men of each maniple held their shields above their heads and closed up until the shields overlapped and formed a roof over the whole body. Under these shell-like covers, resembling huge tortoises, the maniples moved up in succession to the walls.[3] This manner of attack was often resorted to in later times, and was used by Cæsar in the Gallic wars. The name tortoise was also applied by the classical writers to an enclosed battering-ram, from the analogy of the moving in and out of its head; and to a pent house on wheels.

Tall siege towers were in constant use. At the siege of Marseilles in 49 B.C. Cæsar built a stationary tower, 30 ft. square and six storeys in height, under the very walls of the city and in face of a rain of missiles from its engines. The walls of the tower were of brickwork 5 ft. thick. When the lowest storey was built it was covered with a solid fireproof roof which was not secured to the walls but rested upon them like a lid. The eaves projected considerably, and from them screens were hung on all sides, covering all the walls. By means of screws the whole canopy, roof and screens, was now raised to the height of one storey and the workmen proceeded to build the walls of that storey under its protection. This process was repeated in the same manner until the full height of the tower was attained.[4]

Pent houses, called variously *vinæ*, *musculi*, and even *testudines*, were also in frequent use. A pent house was a covered passage provided for the protection of men employed in sapping or undermining a wall; or in filling ditches. It was usually built of timber and covered with raw hides. At Marseilles Cæsar constructed a pent house between his stationary siege tower and a tower of the city he wished to undermine. The pent house (*musculus*) was 60 ft. long, was built of heavy timbers, and had a sloping roof with three coverings. The under covering was of tiles set in mortar, the next of hides, and the outer, to deaden the effect of blows from stone missiles, of fire-

---

[1] *Cæsar: de Bello Civili*, Book II, C. 2.   [2] *Polybius*, XXI, 27.
[3] Ibid. XXVIII, 11.   [4] *Cæsar: de Bello Civili*, Book II, 8, 9.

resisting mattresses. Barrels, filled with resin and tar and set alight, were cast down from the wall on the pent house; but they rolled off the sloping roof and were promptly removed with long poles and forks. Eventually, under the protection of their covered way and the engines working in the siege tower, the sappers were successful and brought down the tower they were undermining.[1]

Mining, while generally effective in reducing a fortress, sometimes failed, and those engaged in it were repulsed with considerable loss. At the siege of Ambracia in 190 B.C. the Romans, with their battering-rams, broke through the wall again and again. But each time a breach was made they were faced with a new wall which had been built up hurriedly behind the old one. As a last resort they took to mining, hoping to carry out the work secretly. For many days they were able to dispose of the excavated earth unobserved, but the dump was eventually seen by the citizens who at once took measures to discover the position of the mines. They dug a trench on the inside and parallel with the wall, lining the side of the trench nearest the wall with very thin plates of brass. By the vibration of these plates at a certain part they were able to locate the mines and by countermining broke into them. They arrived none too soon, for the enemy had not only reached the wall but had dug out and underpinned a long stretch of it. A desperate but indecisive fight ensued and the Ambracians, failing to dislodge the Romans by force, resorted to measures which proved more effective.

They procured a large corn jar, filled it with feathers, placed on the top of the feathers some pieces of burning charcoal, and covered the mouth with a perforated iron lid. They then pierced the bottom of the jar with a hole into which they inserted a tube, passing the tube axially through the jar until it reached the charcoal. The jar was now placed in the tunnel with its mouth towards the Romans, and all the space about it, with the exception of one hole on either side, was sealed up, the side holes being for the pikes they thrust through to prevent the enemy approaching the jar. With a blacksmith's bellows they now blew into the tube on to the burning charcoal, gradually withdrawing the tube as the feathers caught fire. Soon a most nauseating and pungent smoke from the feathers was forced into the mine and the Romans, unable to endure the repugnant fumes, were forced to abandon their works.[2]

---

[1] *Cæsar: de Bello*, Book II, 10.   [2] Hero. From *Polybius*, XXI, 28.

CHAPTER IV

FORTIFICATIONS OF THE ROMAN EMPIRE

THE Romans governed for so long a period, penetrated and occupied their vast empire so thoroughly, and built so substantially, that there still exist throughout Europe, the Levant, and North Africa, extensive remains of their military works. It is only possible here to point out some of the principles of these fortifications and their development, and to describe some examples in Italy, Palestine, Gaul, and Britain.

The camps built by the Romans when laying siege to a city had developed in the course of time into a plan so perfect that it was taken as a model for the permanent camps founded in various parts of the Roman empire, and also, to some extent, for the lay-out of new cities. These camps were rectangular enclosures, surrounded by a rampart or wall and, generally, by one or more ditches. There were usually four gateways, one near the middle of each side; the principal gate being called the porta prætoria. In the centre of the camp, facing the porta prætoria, was the tent of the commanding officer, the prætorium, and running right and left in front of the tent was the via principalis, terminating at one end in the porta principalis dextra and at the other in the porta principalis sinistra. Running from the porta prætoria to the prætorium was the via prætoria, and roughly in line with it behind the tent was often a street leading to the fourth gate, the porta decumana. Sometimes there was still another street, the via quintana, which ran behind the Prætorium, parallel with the via principalis, and occasionally terminated with a gate at either end.

Josephus, describing the Roman camps, said that the walls were strengthened by towers and that engines for throwing stones, darts and arrows were stationed along the defences between the towers. "They also erect four gates, one at every side of the enclosing wall, and those large enough for the entrance of beasts and wide enough for making excursions if occasion should require. They divide the camp within into streets, very conveniently, and place the tents of the commanders in the middle, but in the very midst of all is the general's own tent, in the nature of a temple; insomuch that it appears to be a city built on the sudden, with its market place and place for handicraft trades, and with seats for the officers superior and inferior, where, if any differences arise, their causes are heard and determined. The camp, and all that is in it, is encompassed with a wall round about . . . and if occasion requires a trench is drawn round the whole, whose depth is four cubits, and its breadth equal."[1]

One of the earliest permanent fortifications, of which there are substantial remains, constructed on this principle is that of the City of Aosta in Italy, built by Augustus in 23 B.C. on the site of a Roman camp. This fortress forms a rectangle 793 yards by 624 yards, enclosed by a wall 21 ft. high and defended by square towers at the corners and along the sides. There were four gates, of which two, including the Porta Prætoria, still remain. The Porta Prætoria has three arched entrances, the central a carriageway and the others footways, and each entrance was defended by a portcullis. The gateway is 65 ft. long and encloses a large hall extending across the width of all three arches.

At Rome, A.D. 23, Tiberius built the Castra Prætoria at the north of the city, outside the ancient Servian walls. The Castra Prætoria was a great rectangular fort, 481 yards long by 415 yards wide; it was enclosed by a battlemented wall 12 ft. high and accommodated 10,000 men of the prætorian guard. The men lived in vaulted chambers, some of which were arranged back to back in long rows, running north and south; each row being two storeys in height. Other chambers were built against the inside face of the enclosing wall and probably formed the supports of the original wall walk. When the outer city walls of Aurelian were built, A.D. 271–275, the camp was included within the city and its walls were raised to the height of 24 ft. In 312 Constantine the Great disbanded the prætorium guard, broke up the camp, and pulled down the wall between the camp and the city; but the outer walls, being part of the defences of the city itself, were preserved. These walls still exist, and at the north and east sides of the fort are the remains of some of the chambers.

Probably the finest works of military architecture of about the beginning of our era were those built by the Jews in Palestine under the Herods. On his capture of Jerusalem in 37 B.C. Herod the Great raised the walls of the city and strengthened the fortifications throughout. Thirteen years later, on the site of the present citadel, he built a palatial castle, flanked by three powerful towers, named Hippicus, Phasæl, and Mariamne, after his friend, his brother and his wife respectively. These towers were large square structures battlemented at the top and built of huge blocks of stone; they were solid from the ground to nearly half their height, and above that level were divided into storeys of well-appointed chambers. Phasæl, the largest, rose to a height of about 140 ft.; its upper chambers were designed as magnificent living apartments and contained a bath and every requisite appertaining to a royal palace of the period. Josephus said that this tower in appearance resembled the celebrated Pharus of Alexandra.[2]

[1] Josephus. *The Jewish War*, Book II, C. V, 1–2.
[2] Josephus op. cit., Book V, C. IV, 3–4.

When Titus took and destroyed the city A.D. 70 he spared these three towers on account of their great strength. And, despite the destruction wrought by Julius Severus A.D. 133, and all the other vicissitudes it has undergone, there can be little doubt but that the lower part of one of them still exists in what is now called the Tower of David in the Citadel. A typical stone near the base of this tower measured by the writer was 10 ft. long by 4 ft. high, and had a drafted margin from 4 in. to 5 in. wide.

Herod Agrippa I about A.D. 41 laid the foundation of the outer defence, or third wall of Jerusalem. The wall was 15 ft. thick, and was being constructed in so powerful a manner with enormous blocks of stone that Agrippa, fearing to excite the suspicions of the emperor, suspended the operations. But the work was resumed and completed after his death A.D. 44. The wall was 30 ft. high to the wall walk and was finished with battlements and strengthened at frequent intervals with large square towers. The towers were built of ashlar, they were solid to the height of the wall walk and rose up in storeys of chambers 30 ft. above that level. On the top they supported cisterns for the storage of rain water. Foundations of this wall with its huge blocks of stone were discovered a few years ago.[1] At the north-west of the city, where the third wall turned southward, the Jews built a large octagonal tower, about 110 ft. high, which they called Psephinus.[2]

At the time of the Herods the cities of Palestine throughout were strongly defended and at least one city, Joppa, was protected by two lines of walls.[3] But among the most powerful strongholds in the country was that of Masada, built by Herod the Great about 30 B.C. on the site of a more ancient fortress.

Masada stands on a plateau on the top of a high hill near the west shore of the Dead Sea, and has strong natural defences. The hill is precipitous on all sides, and the fortress on the top is inaccessible except by a narrow path which winds sinuously up on the west. The plateau on the crown is roughly egg-shaped, having its major axis, about a third of a mile long, running north and south. Herod surrounded the whole plateau by a stone wall, about 12 ft. thick by 18 ft. high, and strengthened by thirty-eight towers, 75 ft. high and arranged at intervals along the curtain. On the western side of the enclosure, commanding the path of approach, he built a strong palace or keep; and on the path itself, some distance down the hill from the fortress, he built a tower as an advance post. Within the fortress numerous dwellings and other buildings were erected; and, owing to the scarcity of water, Herod provided an extensive system of rock cut cisterns, which were excavated in various parts of the enclosure and provided the fortress with an abundant water supply.

[1] *The Third Wall of Jerusalem.* E. L. Sukenik and L. A. Mayer, 1930.
[2] Josephus, V, IV, 3.
[3] Ibid. III, VII, 31.

The keep was rectangular in plan and was built on principles at once of strength and of splendour. It was enclosed by lofty walls and had at each corner a powerful tower 90 ft. high. Internally the design was on a sumptuous scale. Cloisters and a great variety of baths were provided, and the walls and floors were covered with slabs of stone of various colours.[1]

The fortress of Masada, both in design and purpose, resembled the mediæval castle in a striking manner. In design, with its rectangular keep placed near the curtain wall of an extensive bailey, it has many mediæval parallels—as at Arques and Falaise in France, and Rochester and Scarborough in England, while in purpose, like many of the castles of the Middle Ages, it was built entirely as an isolated fortress for the defence of one leader and his followers; and that at a period when all other strongholds were cities intended for the protection of large communities of people.

Not knowing what the future might have in store for him, Herod built Masada as a safe and impregnable retreat where he, and such forces as he could collect, would be able to hold out against almost any odds. With this contingency in view he not only made the lavish provision for the water supply noted above, but also laid in vast stores of corn, wine, oil, pulse, dates and other fruits. In storing these commodities Herod's servants must have been in possession of preservative methods since lost; for the provisions when discovered A.D. 70, nearly a hundred years later, were said by Josephus to be quite fresh and good.[2] Herod also laid in great stores of iron, brass and tin, and of weapons of war of all kinds.

On the fall of Jerusalem, A.D. 70, a large body of Jews held Masada for a long time against the forces operating under the procurator Flavius Silva. In order to prevent any escape Silva built a wall all round the foot of the hill on which the fortress stood and established forts, foundations of which still remain, at strategic points. Selecting an eminence immediately opposite the entrance on the west, he built a great mound and set a tall tower, plated with iron, upon the mound. In the tower he placed his siege engines, and from them poured such a hail of stones and darts at the fortress that the position of the Jews on the walls became untenable. Then, unmolested, he was able to prepare the way and bring up a large battering ram against the wall. After some time a breach was made and a portion of the wall completely overthrown.

But the resources of the Jews were not yet exhausted. Experience had shown that a wall built with timber bonding, being more resilient, offered greater resistance to attacks from battering-rams than those of solid stone. In great haste, therefore, they built, inside the breach, a wall constructed of a framework of timber, and an infilling of earth. When the ram was brought against this new obstruction its blows had no other effect than

[1] Josephus, VII, VIII, 3.
[2] Josephus, VII, VIII, 4.

of shaking the materials together and consolidating the whole work. However, the new wall had evidently been built with greater haste than skill; too much timber had been used and too little ramming of the infilling done. For the Romans, failing in their attempt to overthrow it, threw a great number of burning torches at the barrier and, the timber work having caught fire, the wall was destroyed in the flames. Having achieved this result the Romans retired for the night, knowing that the Jews were at their mercy and that there was no escape. The Jews, believing themselves to be faced with butchery and slavery, first killed their wives and children and then destroyed themselves.

Probably the finest works of the later days of the empire are the fortifications of Rome itself. The emperor Aurelian, A.D. 270-275, extended the confines of Rome and surrounded the enlarged city by a powerful wall 12 ft. thick, and about twelve miles in length. Later, probably under Maxentius, 306-312, this wall was raised to about double its original height. In making the addition the old wall walk was covered in and retained as a lofty gallery with loopholes to the field, a second walk, with battlements, being formed above it. As finished the wall was now, in places, 60 ft. in height, it was strengthened at intervals of about one hundred feet with square towers of bold projection, and was pierced by eighteen gates. Repaired and strengthened at later periods, about two-thirds of the wall and nine gates still remain.[1]

Directly their rule was established the Romans proceeded to build in all countries under their control systems of camps, outposts, frontier lines, and cities.

On the German frontier they constructed a line of fortifications between the Rhine and the Danube which extended for a length of about three hundred miles. The *Limes Germanicus*, as this frontier line is called, was probably completed in the early years of the third century, A.D., when previous works of Vespasian, Domitian and Hadrian were consolidated and strengthened. The frontier runs from Rheinbrohi, on the Rhine, to Hienheim, near Regensburg on the Danube, and is of two distinct kind of works. The western portion, known as Pfahlgraben (or pale), consists of an earthen mound and a ditch; and the eastern portion, the Teufelsmauer (Devil's Wall), is a stone wall 4 ft. thick. Both sections are supported at frequent intervals by signal towers and by rectangular stone-built camps, some of the camps standing close to the rampart and others slightly back from it.

The native Gallic towns, as we have seen, were protected by stone walls, of great resisting strength, at the time of the Roman invasion in 58-49 B.C. But on the conclusion of hostilities the Romans built throughout the land their own system of military works. Extensive remains of these works still exist at Fréjus, Nîmes, Autun, Le Mans, Senlis, and elsewhere.

[1] *The City Wall of Imperial Rome.* I. A. Richmond. 1930.

Fréjus, the *Forum Julii* of the Romans, was a colony of the Eighth Legion, founded in 31 B.C. on the site of a more ancient town. It was an important naval station and had a large harbour, now filled up by alluvial deposits. Among the most interesting remains of the fortifications, which date principally from the time of Augustus and are now in a most ruinous condition, are the Porte des Gaules and the Porte de Rome, both built on a similar plan.

The Porte des Gaules is the best preserved. The gateway is set at the back of a large semi-circular forecourt, the entrance to the court being guarded on either side by a powerful round tower. The gateway itself

Fréjus. Plan of the Porte des Gaules.

is flanked by a tower on either side and has three entrances, the central one for carriages and the other two for foot passengers. This is a particularly strong design, since an enemy assaulting the gate in the forecourt would be under attack from four towers as well as from the side walls and would find himself commanded at once in front, on both flanks, and at the rear.

The Porte Auguste at Nîmes dates from 15 B.C. (p. 35). It has four arched entrances, of which the middle pair are carriage ways, each 13 ft. wide, and the other two, footways each 6 ft. 6 in. wide; the carriage ways were protected by portcullises. The gateway is 52 ft. long, and between the outer and inner walls there is a large hall, which extended across the width of both carriage ways. On either side of the entrances there was a large tower, semi-circular in front and square towards the town.

At Autun, where there are considerable remains of Roman walls, are two gateways, the Porte St. André and the Porte d'Arroux, of about the same period as that at Nîmes. Each of these gates has four entrances, arranged similarly to those of the Porte Auguste; but here, the gateways being relatively short, there are no halls behind the entrances. Over the arches in each case there is an arcaded upper storey connecting the wall walk on one side of the gate to that on the other side.

The Porta Nigra at Treves, built later in the Roman occupation, has two entrances, one 12 ft. 10 in. wide, and the other 10 ft. 4 in. wide; each entrance being protected by a portcullis. Within the gate there is a large hall 57 ft. long by 26 ft. wide.

The Roman double gateways, which provided one passage for ingress and another for egress, had great advantages over single gateways from a traffic standpoint; and it is probable that had the mediæval gateways

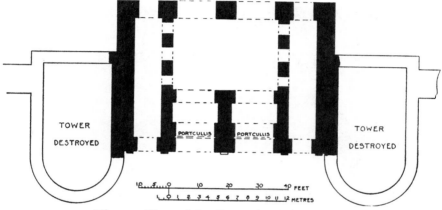

Nîmes. Plan of the Porte Auguste. p. 34.

been built on this design many would have been preserved which have been destroyed as obstructions. But the advantage was confined to times of peace. In an assault, since they multiplied the points requiring defence, they were a source of weakness, and in some cases one half of the gate was subsequently blocked by a stone wall.

At Senlis a considerable portion of the wall and sixteen wall towers, all dating, probably, from the latter part of the Roman period, are still standing, and in some places almost to the full height. The masonry is of stone, bonded at wide intervals with brick lacing courses, two bricks thick. The towers are semi-circular, they are built solid to the height of the wall walk and rise two storeys above that level. The lower storey of each tower is entered by a doorway on the walk, and both storeys are lighted by large round-headed windows (p. 36).

In Britain, following the Roman invasion of A.D. 43, military roads and chains of forts were constructed in various parts of the country, legionary fortresses built at York, Chester, and Caerleon-on-Usk, and towns founded at London, Colchester, Leicester and elsewhere. In A.D. 81 Agricola formed a northern boundary line by building a chain of forts across the isthmus between the Forth and the Clyde.

About A.D. 119 another boundary was drawn further south along the course of an older military road. The new line consisted of a broad flat ditch, about 30 ft. wide and 9 ft. deep, called the Vallum, which with its mounds of upcast on either side extended across the country between the Tyne and the Solway. Of itself the vallum had little or no military value, being fundamentally a line marking the northern boundary of the Roman province. But it was defended by a series of camps, or stations, built on the advance or north side of the boundary and connected by a military road.

A few years later, about 122–125, a continuous wall, known as Hadrian's wall, was built on the north side of the stations, some of the old stations

Senlis. Roman Walls. p. 35.

being abandoned and new ones added. The wall was seventy-three miles long and for the greater part of its length 7 ft. 6 in. thick. It would appear, however, that the thickness as originally designed was to be from 9 ft. to 10 ft. throughout. A portion near Newcastle, about six miles long, is from 9 ft. 1 in. to 9 ft. 7 in. thick, and most of the wall stands upon an unusually broad foundation. But it is evident that the design was modified early in the course of the work. The height, including the parapet, was probably about 20 ft. For long stretches the wall runs along on the top of high ridges with a precipitous fall on the north side, but on lower ground it is defended by a deep ditch with a level terrace, or berm, between the ditch and the wall.

The camps along the line of the wall are designed on the usual Roman model; there are seventeen of them, spaced about four miles apart. Borcovicium, or Housesteads, about midway in the length of the wall, may be taken as an example. This camp has a rectangular plan, rounded at the angles, it is enclosed by a stone wall, 5 ft. thick, backed on the inside by a clay rampart, 15 ft. thick at the base; and is defended at the angles and sides by square towers which project inwards from the walls. The north wall is continuous with the frontier wall itself. There are four gates, each having a double carriage way and a tower on either side. The buildings within the station were all built of stone and were divided by streets into three main sections; the middle section containing the prætorium, other official quarters, and the granaries, and the other sections the blocks of barracks.

At every Roman mile between the camps there is a milecastle, a rectangular building measuring about 60 ft. by 70 ft. internally, projecting inwards

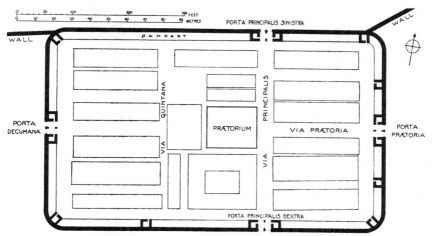

Plan of Borcovicium or Housesteads.

from the wall. And between the milecastles the wall is again divided into three equal sections by turrets, or signal stations, about 13 ft. square internally. A military road, 20 ft. wide, runs along on the south side of the wall connecting all the stations.

Hadrian's wall in itself could have been of no great value against massed attack. For there were no stairways other than those at the camps, milecastles, and signal stations, and therefore no means of ready access to the battlements for troops rushed along the road to repel an attack at an intervening point. But it was a serious obstacle to raiding parties and might even hold in check for a short time large bodies while they were being outflanked by troops issuing from the north gates of the camps. It was also a continuous elevated sentry walk.

In A.D. 143, following the conquest of the Lowlands, the frontier was again moved forward to the line of Agricola's forts between the Forth and Clyde. New forts were built, some of them on old sites, and a continuous rampart, the wall of Antoninus Pius, was thrown across the country from Bridgeness on the Forth to Old Kilpatrick on the Clyde. Bede, drawing a comparison between a wall and a rampart, says: "For a wall is made of stones, but a rampart, with which camps are fortified to repel assaults of enemies, is made of sods, cut out of the earth, and raised above the ground all round like a wall, having in front of it the ditch whence the sods are taken, and strong stakes of wood fixed on its top."[1] Actually this rampart is of sods for about three-quarters of its length, the eastern quarter is of earth, faced with clay. The rampart throughout rested upon a strong stone platform, 14 ft. wide, and, with a steep batter on both sides, rose to the height of about 10 ft.; the walk on the top being about 6 ft. wide. This frontier line was defended by a ditch 40 ft. wide and 12 ft. deep, having a berm between the ditch and the rampart varying in width from 20 ft. to over 60 ft. There are no milecastles but the forts, which project southward from the rampart, are only about two miles apart and therefore much closer together than these on Hadrian's wall.[2]

Colchester, built in the latter half of the first century of our era, and London, in the first half of the second century, may be taken as examples of Roman cities in Britain.

Colchester has a rectangular plan with rounded corners. It was surrounded by a wall about 8 ft. thick backed by a rampart of earth 20 ft. thick, and appears to have had six gates, two on the north, two on the south, one on the east and one on the west. There were also two posterns, one at the east end of the north wall and the other at the south end of the west wall. The wall was built of stone, roughly faced on both sides and tied at intervals with brick lacing courses four bricks thick; it was strengthened at convenient points along the line by square and semi-circular towers. A considerable portion of the wall, four semi-circular wall towers, and the lower parts of the towers of the west or Balkerne Gate still remain. The Balkerne Gate had two carriage ways, separated only by a narrow wall, and two footways, one on the north and the other on the south of the carriage ways. The gate projected out 30 ft. from the face of the wall and was flanked on either side not by a semi-circular or square tower—the more usual forms—but by a quadrant-shaped tower (p. 39). When complete this gate must have presented a most unusual and striking appearance.

Roman London with an area of 330 acres was about three times the size of Colchester. Portions of the land walls—the walls on the north, east

---

[1] Bede. *Ecclesiastical History*, Book I, 5.
[2] *Vide The Archæology of Roman Britain*, by R. G. Collingwood, 1930, Ch. V.

FORTIFICATIONS OF THE ROMAN EMPIRE

and west of the city—which remain have an outside plinth and vary in thickness above the plinth from 7 ft. to 9 ft.; they are built of stone with lacing courses of brick at varying intervals. It is not known whether or not the wall was backed with an earthen rampart. All the gates are destroyed, but the foundations of one of them at Newgate have been discovered. Newgate had a double gateway, each passage being about 12 ft. wide, and was flanked on either side by a square tower. The wall towers, some of which were solid, probably to the height of the wall walk, and some hollow, and the whole of the river wall, are believed to be later Roman additions.

In the construction of their military works as well as in the erection of public and private buildings there can be little doubt but that the Romans adopted the normal practice of employing native talent and labour. Speaking of the works of Agricola Tacitus says that he incited the natives "to erect temples, courts of justice, and dwellinghouses . . . He was also attentive

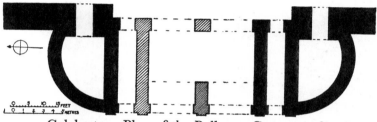

Colchester. Plan of the Balkerne Gate. p. 38.

to provide a liberal education for the sons of their chieftains, preferring the natural genius of the Britons to the attainments of the Gauls."[1]

During the last century of their occupation of Britain the attention of the Romans was diverted from their northern frontier to their southern coast line. From about the end of the third century A.D. the raids of the Saxons on the eastern and southern shores of England became more and more frequent and dangerous. The menace was so serious that the Romans built a series of forts, called forts of the Saxon shore, on the east and southeast coast. Extensive remains of these forts still exist, especially at Burgh, Reculver, Richborough, Pevensey and Porchester; that at Richborough occupying the site of a defensive work of A.D. 43.

Reculver, Richborough and Porchester are all rectangular forts. At Reculver the walls are 10 ft. thick at the base and rise in off-sets to 8 ft. thick at the top. The walls of Richborough Castle, which are of stone with brick lacing courses, are 11 ft. thick and rise in places to a height of 25 ft.; they are strengthened by round towers at the angles and square towers at the sides. A remaining gateway on the west has a single entry 11 ft.

[1] Tacitus. *Life of Agricola,* Ch. 21.

wide, flanked by square towers. Pevensey is oval shaped; its walls still stand in places to the height of the wall walk, 28 ft. above the ground. Both Pevensey and Porchester were fortified by the Normans, who in each case built a castle within the Roman walls.

In these forts the wall towers now project on the outside of the curtain, instead of on the inside as in the camps on Hadrian's wall. This disposition gave greater range to the ballistæ mounted on the towers, especially at the corners. At York, the defences of which were reconstructed at this period, a large polygonal tower, or bastion, was built at each of the west and south corners of the curtain. The lower part of that at the west corner, the Multangular tower, now surmounted by a mediæval storey, still remains (p. 8). Here the projection is so great that an engine mounted on the tower would have a lateral sweep, outside the walls, of more than three-quarters of a circle. Internally the tower is divided by a cross-wall, built to support the upper floor and the engine mounted upon it.

The last phase of Roman fortification in Britain was the extension of the Saxon shore defences northwards. In the last half of the fourth century the Romans built a line of signal stations on the headlands of the Yorkshire coast. Remains of these stations have been found between Huntcliff near Saltburn and Filey. They were square towers, built of stone, and were defended by an outer wall and a ditch, the outer wall having a bastion at each corner.

SIEGE ENGINES

The siege engines of this period were very powerful and were used in large numbers. At the siege of Jotapha, A.D. 68, Vespasian brought up a battery of one hundred and sixty engines against the walls of the city.[1] In his description of Vespasian's army in marching order Josephus shows how the engines were transported from place to place. He says that the engines for sieges, and other warlike machines of that nature, were carried by mules; evidently in sections which could be reassembled in position.[2]

At the siege of Jerusalem, A.D. 70, Titus used engines which threw stones of 1 cwt., striking with tremendous force objects a quarter of a mile away.[3] At the same siege Titus built three great siege towers, 60 ft. high, and covered with iron plates.[4]

---

[1] Josephus op. cit., III, VII, 9.
[2] Josephus op. cit. III, VI. 2.
[3] Josephus op. cit. V, VI, 3.
[4] Josephus op. cit. V, VII, 2.

CHAPTER V

BYZANTINE FORTIFICATIONS FROM THE FIFTH CENTURY TO THE TENTH CENTURY

FOLLOWING the withdrawal of the Roman legions from the outposts of the Western Empire and the fall of Rome, the Western nations were left to defend themselves against their barbarous foes as best they might. Henceforth for many centuries they were involved in desperate strife of a kind which allowed little or no opportunity for the development of scientific fortification. It was in the Byzantine Empire that progress in this direction occurred.

Here, while the Persian frontier was maintained on the East, the advance of the Goths and Huns was held in check on the north and west. To ensure the protection of Constantinople from the onslaughts of the latter, in 413

Constantinople. The Land Walls.

a great wall, strengthened every sixty yards by powerful towers, was built across the west, the land side, of the promontory on which the city stands. Having suffered severely from earthquake this wall was repaired in 447, when an outer wall was added and a wide moat was dug before the outer wall. The walls are built of stone with a concrete core and are bonded with brick lacing courses five bricks thick. The inner wall is 15 ft. 6 in. thick at the base and rises to a great height. The outer wall, 6 ft. 6 in. thick, is constructed with a continuous series of internal arches. There are two wide terraces, one between the walls and the other between the outer wall and the moat. This noble fortification with its triple lines of defence, the inner and outer walls and the moat, repaired and in some places altered by later emperors, was a powerful bulwark of defence against the attacks of successive invaders; and even now, shattered by earthquake and neglected,

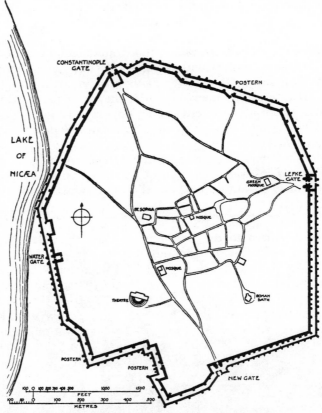

Plan of Nicæa. After Tercier and Pullan.

it is one of the most imposing and inspiring works of its kind in existence.

The fortifications of Nicæa in Asia Minor, though incorporating much work of earlier dates, are largely of about the middle of the fifth century. They have been ascribed, recently, to Justinian. But, apart from the fact that they resemble the land walls of Constantinople both in structure and disposition, it is not without significance that Procopius in dealing with the subject at length does not mention the fortifications in his descriptions of

BYZANTINE FORTIFICATIONS

Justinian's works at Nicæa.[1] Nicæa (p. 42) is surrounded by a double line of walls, the inner wall of great thickness and height, and the outer lower and less substantial. Both walls are built of stone with brick lacing courses and are strengthened by towers, placed at frequent intervals; the wall towers being so placed that those in one wall stand opposite to an interval in the other wall.

A large residential tower on the south side of the city, higher and more powerful than the others, must have been a kind of keep, or donjon, in the defences. It was against this tower, then occupied by the Sultan's wife, that the Crusaders concentrated their assault in the siege of 1097. And it was not until attack after attack had been made upon it that the tower was eventually brought down, and only by means of undermining its walls. There are four gateways and three posterns. The walls of Nicæa have withstood many attacks; they repelled the crusaders again and again during the memorable siege of 1097, and are still in good state of preservation.

Of the military works on the confines of the Byzantine empire those of the fortress of Babylon of Egypt, now called Old Cairo, call for special attention. This fortress, rebuilt in its present form during the latter part of the fourth or early part of the fifth century, stands on the right bank of the Nile at the head of the delta, and commanded not only the river at this strategic point but also the passage across the river of the great caravan route from North Africa to the East (p. 44). On the side of the fortress towards the Nile and directly commanding the passage are two round towers each 90 ft. in diameter and placed 66 ft. apart. These towers are of exceptional interest, not only on account of their unusual design and great size, but also because they stand close together and are built to correspond with each other. They are built of small squared stone with brick lacing courses, all bonded in with the curtain walls, and stand upon foundations of large blocks of stone. Each tower consists of two concentric walls, spaced 15 ft. apart, and of eight radial ribs which connect the two walls and divide the intervening space into eight equal compartments. This is a particularly powerful and scientific method of construction. For while the combined walls have the effective strength of a single wall 28 ft. thick, there is a great saving of material, and the tower is provided with spacious wall chambers in addition to a large central apartment. The ribs are of great strength and are true radians, offering great resistance to attacks on the tower. The compartments themselves are a source of strength rather than weakness, since if one of them is broken into by a siege engine, the damage is localized and its repair made relatively easy. One of the compartments in each tower, that immediately inside the curtain wall and facing the tower opposite, is occupied by a newel stairway.[2]

---

[1] Procopius: *Of the Buildings of Justinian*, Book **V**.
[2] *Vide* "Babylon of Egypt" by the Author. British Archaeological Journal, 1937.

About A.D. 500 Anastasius, in his efforts to put a further check to attacks on the capital from the west, built the "Long Wall" which, from a point on the Marmara forty miles west of Constantinople, stretched northward across the land to the Black Sea. The wall was about fifty miles in length

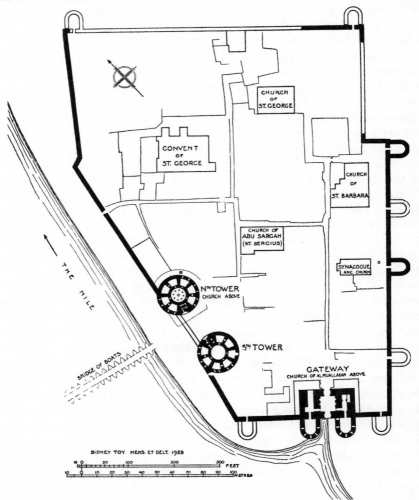

Old Cairo. Plan of the Fortress of Babylon in Egypt. p. 43.

and was strengthened by numerous towers. Anastasius also fortified the city of Dara in Asia Minor, on the Persian frontier of the empire.[1]

During the reign of Justinian great strides were made in military architecture. The Byzantine fortifications of this period are among the greatest

[1] Evagrius: *Ecclesiastical Hist.*, Book III, 37, 38. Procopius: *Of the Buildings,* Book IV.

works of military engineering of all time. Justinian not only strengthened existing fortresses throughout his vast empire but built numerous new ones. During his brilliant reign of thirty-eight years—527 to 565—he rebuilt the fortifications of many large cities, strengthened those of others, and built or repaired numerous forts; his military works numbering about 700 in all. The building activity of the Eastern empire at this period is without parallel. And it was largely to the scientific and powerful works then built, many of which still exist, that the mediæval military engineer, both Christian and Saracen, owed his inspiration.

At Dara, about 140 miles N.W. of Mosul, the fortifications of Anastasius, having been hastily constructed, proved to be weak and inadequate. Justinian repaired and strengthened them. He increased the height of the curtain wall by building a vaulted gallery, with loopholes to the field, upon the wall walk along the whole line of the fortifications, and by making another wall walk with battlements upon the gallery, so that there were now two fighting lines, one above the other. A second curtain wall was built outside the first, leaving a space of fifty feet between the two walls. The inner wall was 30 ft. thick at the base, and, diminishing in thickness, rose to the height of 60 ft.; the towers in this wall were 100 ft. high. The outer wall was smaller but was also provided with towers, so placed that they stood opposite to an interval in the inner wall.

One of the towers of the inner wall, called the watch tower, appears to have been of greater importance than the others and a kind of keep; this tower was entirely rebuilt by Justinian.

In all his works Justinian gave great attention to the water supply of his fortresses. Dara obtained its supply from a stream which entered and passed out of the city through conduits in the wall, strongly guarded by iron bars. At the inlet the stream was protected by the mountainous nature of the ground in front, but beyond the city the stream was a source of weakness, since it provided an abundant supply to an enemy encamped close to the walls. By the fortunate discovery of an underground passage, which could be entered by a shaft within the city and had its outlet many miles away, it was possible in time of siege to divert the stream into this subterranean passage and so cut off all the enemy's supply.

At Edessa water was obtained from a river which ran through the city and which in time of flood caused great loss of life and destruction of buildings. Justinian, by cutting a deep channel through high ground on one bank, and building a wall of enormous stones on the other, diverted the main course of the river round the walls. In this manner he not only saved the city from floods but also provided a moat for the hitherto unprotected walls. The part of the stream still allowed to follow the old course was carried through the city in a channel of masonry.

At Theodosiopolis the walls were heightened by building a gallery over the existing battlements and a second line of battlements over the gallery, as was done at Dara. Here the wall towers were so strengthened that each of them became virtually a keep in itself and could be held against an enemy.

The above works were on the Persian frontier of the empire. The other frontiers and outposts, from Egypt to the Danube, received equal attention.

Following the conquest of the Vandals by Belisarius, A.D. 533–534, fortifications were built throughout the newly acquired territories in North Africa. Here it was particularly necessary to guard against the forces of revolt within as well as against the activities of enemies without the borders, and the military works were of great variety. They included fortified towns, such as Guelma, Thélepte, and Bagai; open towns with fortified citadels, as Haidra, Mdauroch, and Timgad; and isolated forts, as Lemsa and Aïn Tounga. The first two classes were at once military stations and places of refuge in times of trouble for the civil populations of the neighbourhood, or the city. The last were isolated castles occupying strategic positions, keeping watch over a plain, commanding an important valley, or guarding a pass. There were also outposts keeping watch at the borders. The Greeks had developed a scientific code of signalling from beacons, by means of which information as to the composition and character as well as of the numbers of an invading force could be signalled from station to station. Scattered throughout Algeria and Tunisia there are large numbers of these Byzantine fortifications, dating principally from the sixth century, many of them still in excellent state of preservation.[1]

These fortresses, though often rectangular, differ greatly in plan; some of them having a very irregular outline. The curtain walls are from 7 ft. 6 in. to 9 ft. thick, and, where the upper parts still exist, are from 26 ft. to 32 ft. high; they are often arcaded on the inner face like the ancient walls of Rhodes, p. 13. The wall walks are protected on the side towards the fortress by a low wall and on the other side by crenellations about 5 ft. high; they are approached from the courtyard either by flights of stone steps, built in convenient positions against the walls, or by stairways in the towers. Sometimes greater width is given to the walk by corbelling out the upper part of the inner side of the wall.

Towers of varying shapes but generally square project boldly out from the corners and sides of the curtain. They are usually of two stages, the basement opening to the courtyard and the upper stage to the wall walk on the curtain; the division between the stages being either a stone vault or a timber floor. The battlements are reached by internal stairways. Sometimes the towers have no doorways to the curtain wall and are capable of being held independently.

The gateways are always defended by one and sometimes by two towers,

[1] *Vide L'Afrique Byzantine*, by C. Diehl. Paris, 1896.

BYZANTINE FORTIFICATIONS

one on either side. Often they pass through the lower stage of a tower, going either straight through, or, entering at one of the lateral walls and issuing at the inner wall, take a right-angled turn within the tower.

As at Nicæa and Dara, there is often one tower which is larger, stronger and better fortified than the others. It occupies a position either at a strategic point behind and completely isolated from the curtain, or is placed in the curtain itself, at the highest point or at a point particularly exposed to attack. This tower was in all essentials of its design and purpose the prototype of the rectangular keep or donjon of later days. It was the strongest building in the fortress, was capable of offering independent resistance, and was the place where the last stand was made.

Bagai, Algeria, is a good example of a fortified town. Here three of the wall towers, projecting boldly out from salient angles of the curtain, are round, while the other towers are square. Occupying a commanding position on the north side of the town there is a citadel and within the citadel a powerful donjon about 85 ft. square. Haidra, Tunisia, and Timgad, Algeria, are well preserved citadels, each guarding an open town; in the former the curtain walls are built with internal arcades.

The plan of the citadel at Timgad forms a regular rectangle. There are towers at the angles and in the middle of each side, all of them square and all projecting entirely on the outside of the curtain. The principal

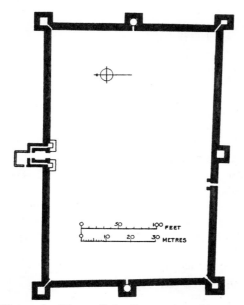

Timgad. Plan of the Byzantine Citadel.
After Saladin.

gateway passes through a particularly large tower in the middle of the north wall and was defended by a barbican; the barbican, which was entered through a lateral wall, being of slightly later date than the gateway, but still of the Byzantine period in North Africa, 534–709.

The inner door of the gateway was flanked on either side by a tower, now destroyed, and from each of these towers a mural passage runs through a lateral wall of the gateway to the outer door, so that an enemy who had carried the first door and was held up by the second could be attacked from the rear by men issuing from the inner towers through the passages. In the south wall of the citadel there is a postern defended by the middle tower on that side.

Probably the best preserved of the isolated forts is that of Lemsa, Tunisia, a rectangular castle with corner towers. Three of the walls stand to their full height, from 26 ft. to 32 ft., and retain their crenellations. Aïn Tounga, guarding a pass in Tunisia, is of trapezoidal plan and has a tower at each corner, one of which, a strong rectangular building, about 36 ft. by 40 ft., is much larger than the others and projects entirely on the outside of the curtain. The gateway passes through a tower near the middle of the south side of the fort, entering at one of the lateral walls and taking a right-angled turn in passing through.

The castle of Gastal, Algeria, forms a single rectangle with a round tower of bold projection at each corner and a square tower in the middle of one side.

In addition to the above types there are also numerous smaller forts scattered throughout Algeria and Tunisia, here guarding a narrow defile, there the approach to a village or important agricultural centre. These forts are plain square or rectangular buildings, having no corner turrets and only one gateway.

The fortifications of the Roman emperors on the Danube consisted mostly of single towers posted along the river, principally on the right bank. These towers which had been destroyed by Attila in 446, were rebuilt by Justinian in much stronger form, and many others were added at suitable points.

At Episcopa, near Silivri on the Sea of Marmara, Justinian built entirely new fortifications, designed by one Theodorus Silentiarius "a very clever man."[1] Here the wall towers were of such bold projection that they commanded every point of the curtain between them. The gates were not designed in the usual manner between two towers in the wall. But each gate was placed at a point where the curtain wall took a short right-angled break inwards. The gateway was at the side through the short wall of the break and was therefore in large measure hidden from view. An enemy attacking one of the gates found himself in an angle of the fortifications, exposed to fire from the curtain wall on his flank as well as from the gate directly in front of him.

[1] Procopius: *Of the Buildings*, IV, 8.

In fortifying the Dardanelles Justinian built a strong fortress at Elæus near Cape Hellas at the western entrance to the Straits. Here he constructed a wall of great width and height and dug a deep ditch in front of it. Upon the wall were raised two storeys of battlements, the lower one of which was vaulted and contained chambers for the garrison.

When Belisarius repaired the fortifications of Rome, after his capture of the city in 536, one of the improvements he made was the addition of wing walls to the battlements; short screens projecting inwards from the left side of the merlons, as at Pompei (p. 24). Remains of these merlons are still to be seen in the fortifications.[1] Belisarius also surrounded the wall with a moat.[2]

The rise of Moslem power in the seventh century put the fortifications of the eastern empire to a severe test. Between 637 and 655 the Saracens had conquered Syria, Egypt and Persia; their fleets had swept the Levant and taken possession of Cyprus and Rhodes; and in 668 they appeared before the walls of Constantinople and laid siege to the city. The siege lasted until 675, when the Saracens were beaten off with great loss; but it would appear that the Byzantines owed their preservation to their use of that powerful weapon "Greek Fire" as much as to the strength of their fortifications.

Greek Fire, the precise composition of which is unknown, though sulphur, naptha and quicklime seem to have been major ingredients, was a terrible weapon either on sea or land. On sea it was blown out of large copper tubes erected on the prows of vessels selected for that purpose. Water would not quench it but rather spread the fire hither and thither. On land it was made up into tubes, phials, or pots and either cast by hand or projected from engines at the end of arrows and bolts.

In 716 Constantinople was again invested by the Saracens who attacked the city both by sea and land. This siege lasted thirteen months and ended in the utter destruction of the Saracen fleet by the fire ships of the emperor, and the rout of their army before the land walls by his allies.

From this period to its fall the empire was constantly assailed from all quarters; but the traditions of skilful military architecture were well sustained throughout, particularly under the early Isaurians and the Comneni.

---

[1] *The City Wall of Imperial Rome.* I. A. Richmond, Oxford, 1930. p. 72.
[2] Procopius: *History of the Wars,* V, XIV, 15.

CHAPTER VI

FORTIFICATIONS OF WESTERN EUROPE FROM THE FIFTH CENTURY TO THE ELEVENTH CENTURY

MEANWHILE the countries of Western Europe had little or no respite for the establishment of permanent fortifications. Spain was overrun by the Saracens, who, but for their decisive defeat at Poitiers in 732, would have overrun France also. Italy, Germany, France, and England were all engaged in continuous and desperate struggle for existence; either with other sections of their own peoples, or against wave after wave of barbarous invaders. From the fifth century to the tenth century the military architecture of those countries consisted largely in the repair of existing Roman fortifications or the building of others on the same plan; even the military castles of Charlemagne were designed on the Roman model.

Contemporary chroniclers confine their notices of buildings mainly to those of ecclesiastical or civil character, but here and there references to fortresses occur. In a poem written at the end of the fifth century, Fortunat, Bishop of Poitiers, describes a castle built on a precipitous eminence on the banks of the Moselle; the foot of the hill being washed by the river on one side and by a stream on the other. The curtain wall was strengthened by thirty wall towers, and a tower containing a chapel and armed with ballistæ guarded the approach. The keep (*aula*) stood on the summit of the hill and was of considerable size and magnificence.[1] Other references occur, but they are mostly vague in character, and, having regard to the paucity of remains and the facility with which the Normans overran Northern and Western France in the ninth century, the military works, in France at least, at that period could not have been of any great strength. Indeed, so little faith was placed in fortifications by Charlemagne and his immediate successors, that permission was given to some bishops to pull down the walls of their cities and use the material in building their cathedrals and churches.

But the raids of the Normans demanded that France should put itself in a state of defence. In 862 Charles the Bald, King of France, ordered the construction of fortresses at all points to resist the invaders. The response to this order was immediate; bishops began to repair and rebuild the walls of their cities and counts to build private castles. The multiplication of private strongholds became so great a menace to the authority of the crown that in 864 Charles issued the Edict of Pistes, ordering the destruction of all fortresses

[1] V. H. C. *Fortunati* - - - - *Opera Omnia*. M. A. Luchi. Rome, 1786. Ch. XII, p. 95.

which had been built without royal licence. But by a further edict of 869 instructions were issued for the fortification of all towns between the Loire and the Seine.

The introduction of the feudal system in France in the tenth century gave additional stimulus to Castle building.

In England, following the departure of the Roman legions, the inhabitants continued to defend themselves behind the Roman fortification at London, Colchester, Leicester, Lincoln, York, Chester, Pevensey and elsewhere. But the successive inroads of Jutes, Saxons, Angles, and Danes; the continuous and desperate strife between these peoples after their arrival and penetration of the country; and the constant passing of power into the hands of fresh nations of rovers, were all factors inimical to the establishment of fixed defences. The conflict was ceaseless; but it was a war rather of pitched battles than of sieges. Both the Saxons and Danes constructed earth works here and there throughout the country. But it was not until after the Peace of Chippenham in 879 that Alfred the Great and following him his son and daughter were able to turn their attention to systematic fortification.

It was essential for the progress and development of the people that each community should be provided with adequate defence and be secure at least from sudden attack. To this end burhs, or fortified towns, were built in suitable positions in various parts of the country. From 910 to 924 Edward the Elder and his sister Ethelfleda, Lady of the Mercians, built over twenty such fortified towns, including Hertford, Tamworth, Stafford, Warwick and Towcester. They also repaired and strengthened existing defences, including those at Huntingdon and Colchester.

Of the character of these fortifications we have very little reliable evidence. The illustrations in Anglo-Saxon manuscripts which show towns surrounded by stone walls and towers are not helpful in this respect, because the illustrations themselves have been deliberately and meticulously copied from more ancient types and can have no local application. The *Psalter, Harley*. M.S., 603, for example, dating about A.D. 1000, contains illustrations which have been reproduced again and again as illustrative of Saxon buildings and life. But these drawings were closely copied from those in a manuscript of the ninth century, now in the Library at Utrecht and known as the Utrecht Psalter; they again were copied from earlier classical types, and this is a typical example. The Anglo-Saxon Chronicle states tersely in respect to the towns that they were built or repaired, as the case may be; except in respect to Towcester, which, the chronicle says, was encompassed with a stone wall.[1] Florence of Worcester speaks of old towns, constructed of stone, which by orders of Alfred the Great were moved from their old sites and handsomely rebuilt in more fitting places.[2] Therefore it may reasonably be inferred that

[1] *Anglo-Saxon Chronicle*, Anno. 921.
[2] *Florence of Worcester*, Anno. 887.

where stone could be procured the fortifications of this period were of masonry. But where wood was plentiful and stone scarce they were probably of timber.

With the advent of the Normans at the court of Edward the Confessor in the middle of the eleventh century a type of fortification, much employed in Normandy, was introduced into England. In France, as early as the ninth century, as we have seen, the feudal lords were building not fortified towns but private castles. By the middle of the eleventh century a special form of fortification had developed now known as the motte and bailey type and consisting of a motte, or mound, varying from 10 ft. to 100 ft. in height and from 100 ft. to 300 ft. in diameter, and of one or more wards or baileys; the motte and baileys being surrounded by ditches. Fortifications of this type are found in Germany, Italy, and Denmark, but they are most profuse in Normandy and in Norman England.

The mounds are of three kinds; natural hillocks, partly natural and partly artificial, or wholly artificial, according to the nature of the sites they occupy. And it should be noted here that mounds are often described by writers as artificial before any investigation as to their composition has been made. The mound at Launceston Castle, hitherto referred to as artificial, was found, on excavation being made on the summit a few years ago, to be a natural hillock throughout.

Occasionally the mound stood alone surrounded by its ditch, but normally it was protected by one, two, or more outworks or baileys; the baileys being arranged in such order, dictated by the character of the site, as would best defend the keep on the mound. If there were two baileys, in some cases they were in line in front of the mound, in others on either side of it, in others again side by side in front of the mound. A ditch was carried all round the fortification and also between the mound and the baileys and between the baileys themselves. In some cases the mound stands entirely within the bailey, as at Bramber in Sussex and Skenfrith in Monmouthshire. At Arundel the mound stands exposed on one side in the centre of a long and relatively narrow bailey. At Lewes and Lincoln there are two mounds, and at Heddingham and Old Basing there are no mounds at all. (p. 53).

Castles such as these, placed in strategic positions, while being a source of danger to a weak ruler when held by truculent and disaffected lords, were of great value to a powerful king, such as William the Conqueror, when held by his trusted vassals. A few castles were built in England during the reign of Edward the Confessor, as at Ewias Harold and Richards Castle, both in Herefordshire; but following the Norman Conquest they were spread profusely throughout the country.

If the mound was artificial, then the defences on its summit must have been of timber, and such a fortification is depicted on the Bayeux tapestry and described by Jean de Colmieu, writing about 1130.

Speaking of the flat open country south-east of Calais Colmieu says: "It is the custom of the nobles of that neighbourhood to make a mound of earth as high as they can and dig a ditch about it as wide and deep as possible. The space on the top of the mound is enclosed by a palisade of very strong hewn logs, strengthened at intervals by as many towers as their means can provide. Inside the enclosure is a citadel, or keep, which commands the whole circuit of the defences. The entrance to the fortress is by means of a

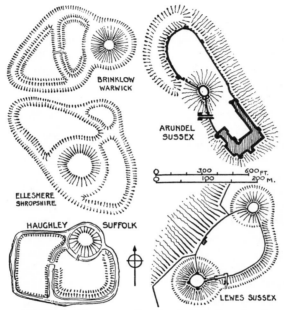

Plans of Early Norman Castles in England. p. 52.

bridge, which, rising from the outer side of the moat and supported on posts as it ascends, reaches to the top of the mound."[1]

Another chronicler, Lambert d'Ardres, writing of a wonderful timber keep built in the eleventh century in the same region, describes an elaborate structure of many storeys and rooms.[2]

In districts where conditions were favourable the keeps were built of stone. The keep of the castle of Brionne in Normandy, built about 1045, was of stone, as described by William of Poitiers, writing in the same century.[3]

The keep at Langeais on the Loire, built at the end of the tenth or beginning of the eleventh century, is of stone. Of the Norman castles re-

---

[1] Jean de Colmieu. *Vie de Jean de Warneton.* Eveque de Téroianne.
[2] *Lambert d Ardres,* C. CXXVIII.
[3] Duchesne Hist. Norman Script. Antiq., 1619, p. 180.

presented on the Bayeux tapestry, one is clearly built of stone, and stands on the ground, while two others are represented much in the same manner as is Westminster Abbey, known to be of stone. In some cases, as at Corfe, though the keep was of stone, the baileys were defended for many years by stockades. There is no evidence that elaborate timber keeps, such as that described by Lambert d'Ardes, were ever built on English soil.

At Durham Castle, built about 1072, the keep on the mound was rebuilt or remodelled in the fourteenth century and again largely rebuilt in 1838–40. It has been stated frequently that the original keep was of timber, but there is no reliable evidence to that effect. The shape of the existing structure is similar to that of many shell keeps built in the eleventh and twelfth centuries, as at Tickhill and Lincoln; while it bears no sort of resemblance to keeps known to have been built in the fourteenth century, as that at Dudley castle. Such parts of the eleventh century curtain at Durham as remain, including that portion running from the chapel to the top of the mound, are strongly built of stone; and there is every reason to believe that the original outer wall of the keep, parts of which are probably incorporated in the existing structure, was of stone also. Prior Laurence's rhapsody on Durham castle, written about 1150,[1] is much more reasonably read as a description of a stone than of a timber building. And it is very improbable that while during the twelfth and thirteenth centuries ponderous and elegant buildings were being raised on the north and west sides of the bailey the opulent bishops of Durham would have been satisfied with a timber keep, their last line of defence, until the middle of the fourteenth century.

The perishable character of timber and its liability to destruction by fire rendered it an unsuitable material for permanent fortification. From the earliest times fire has been one of the principal weapons of offence, and although even stone buildings are not proof against its attacks the damage done to them is partial and often reparable. With timber the destruction is total. The Normans must have been well aware of this fact, and the paucity of remains of castles of the early Norman period may be due in some measure to the pillage of stone for building purposes. At Topcliffe, Yorkshire, every stone has vanished from the site of what was the principal castle of the Percies for six centuries, and of the stone keep on the mound of Richard's castle, Herefordshire, all but a small fragment has disappeared. In some cases masonry may still lie buried beneath the soil. In 1929, on a hill site at Lydney, Gloucestershire, hitherto considered an earthwork, the foundations of a stone-built castle, dating from the twelfth century, were brought to light,[2] and the existence of still earlier masonry, hidden on some other sites, is not improbable.

Timber, however, was of great value for temporary camps and forts which

[1] Surtees Society. Vol. LXX, 1878.
[2] *The Antiquaries Journal*, Vol. XI, 1931, p. 240.

Launceston Castle. The Keep from the East.   p. 59.

Tamworth Castle. From the South-East.   p. 60.

Restormel Castle. The Keep from the West. p. 60.

Restormel Castle. Interior of the Keep, looking towards the gateway. p. 60.

had to be constructed in haste. The Norman castles in England were founded to overawe and govern the districts in which they were built, and a large number of them must have been constructed hastily with such materials as were available. Many of them only served a temporary purpose, and were subsequently abandoned. A large number of them, having regard to the facility with which they were destroyed, as at York, where charred remains of wood have been found on the artificial mound, must have been of timber. Many of the unlicensed or "adulterine" castles raised in haste in England during Stephen's reign and destroyed after the treaty of 1153 were also, probably, of this material.

But where the mounds were natural hillocks and where stone was in abundance and timber scarce, as at Launceston, Totnes, and Corfe, there can be no doubt but that their flat summits were defended by a stone building, known as a shell keep, from the first. It is probable also that some of the shell keeps, which stand upon partly natural and partly artificial mounds, were also built of stone from the first, the foundations going down to the virgin soil. At Skenfrith, where there is a round keep of the late twelfth century standing on an artificial mound, the foundations of the keep are carried down through the mound to the natural soil beneath.

There is a large number of shell keeps in England, some of them dating probably from the eleventh and others from the twelfth century. Totnes,

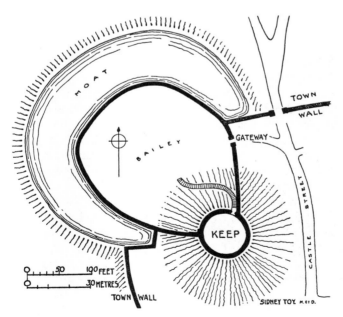

Plan of Totnes Castle. p. 58.

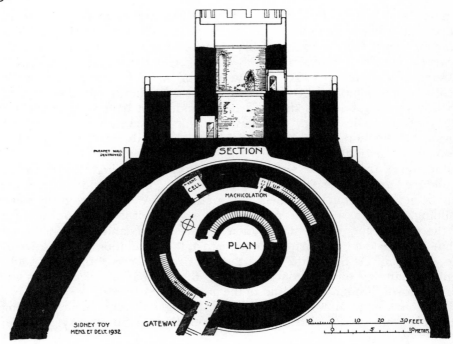

Launceston Castle. The Keep. p. 59.

Launceston, Trematon, Tamworth, Restormel, and Arundel, are examples. Among examples on the continent are those at Gisors in Normandy and Pfeffengen, Switzerland. The living rooms of these keeps were built around and against the inside of the wall; and either occupied the whole space, the roof timbers springing from a central pillar, or they enclosed a central courtyard. In many cases there was a square tower on one side of the shell, built either against or astride of the wall; the tower itself then becoming the keep or stronghold of the fortifications—as at Tamworth, Arundel, and Gisors.

Totnes, Launceston and Trematon, all dating probably from the eleventh century, each consists of a shell keep and a single bailey enclosed by a wall; the keep standing upon a high mound or hillock. At Totnes (p. 57) the keep is a simple irregularly-shaped round shell, 6 ft. 6 in. thick and 15 ft. high; the internal diameter being about 70 ft. Here what internal buildings may have existed have been destroyed. Two stairways, built in the thickness of the wall, rise to the wall walk, which is carried all round the keep and is defended by an embattled parapet. A doorway on one side of the keep leads through a mural passage to a latrine. The curtain wall of the bailey is carried up the slopes of the mound to join the keep on either side. Totnes castle stands above and to the north-west of the town of Totnes, its mound jutting into the

FORTIFICATIONS OF WESTERN EUROPE 59

town and its bailey projecting beyond it. The portion projecting beyond the town is still protected by a wet ditch, but the ditch formerly round the foot of the mound towards the town has been filled in.

The keep at Launceston (pp. 55, 58) is composed of an ovoid-shaped shell and a round tower, built at a later date, probably about 1240, inside the shell. Here the shell, though smaller internally, is much more substantial than that at Totnes. It is 12 ft. thick and 30 ft. high, and has a deep battered plinth crowned by a round moulding. The wall walk was reached by two mural stairways, one near the gateway and one on the opposite side of the keep. A mural chamber with a barrel vault and having a ventilating shaft in one corner was probably a prison cell. The keep is approached up the steep mound, which investigation has proved to be a natural hillock, by a long flight of steps, formerly flanked by walls and covered in by a roof. The foot of the stairway was guarded by a round tower, and at the head stood the round-headed entrance to the keep which was later protected by a portcullis. The transformation of this building from a mere shell to a powerful keep was probably the work of Richard, Earl of Cornwall and titular king of the Romans. The work included the construction of the inner tower and of an embattled walk at the foot of the shell. Considering its commanding position, its three lines of defence, and its magnificent middle platform, this keep when complete must have been amongst the most formidable in England.[1]

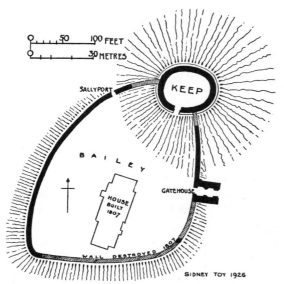

Plan of Trematon Castle. p. 60.

Vide "The Round Castles of Cornwall" by the Author. Archæologia, Vol. LXXXIII. 1933.

The keep at Trematon is ovoid in shape and measures internally 71 ft. 6 in. by 57 ft. Here the internal buildings, traces of which still exist, were ranged along the inside face of the shell. At present there are no stairways to the wall walk. But the stairways must have been associated with the internal buildings, and were destroyed when these buildings were swept away in the eighteenth century. (p. 59).

Tamworth castle is probably another example of late eleventh-century work, at least in respect to the lower part of the walls, the upper part having been either refaced or rebuilt later. It has a multangular shaped shell, about 106 ft. diameter internally, and a square tower on the east which projects slightly beyond the outside of the shell. (p. 55). Here the wall walk was reached by two mural stairways constructed in the thickness of the shell, as at Totnes and Launceston. On the east side of the mound on which the castle stands there is a deep moat, and the castle is reached by means of a stone causeway which is thrown across the moat from bank to bank. This causeway is built of herringbone work and doubtless dates from the eleventh century. (p. 63).

In herringbone masonry the material comprising each course, flat stones or bricks, is laid, not horizontally, but tilted up about 45 degrees. Bond is obtained by tilting the stones in each succeeding course in the reverse direction to that below it, and sometimes further by the insertion of one horizontal line between the courses—as at Tamworth. Two courses of such work present much the appearance of the bones of a herring. This method of construction is not of itself conclusive evidence of any particular period. It was used by the Romans and the Saxons, and was employed extensively by the Normans; it was used in Switzerland in thirteenth century work. Even within the last hundred years it has been adopted in places so far remote from each other as the shores of the Bosporus and the coasts of Cornwall. With the Normans it was often associated with ordinary masonry, and at Colchester, while the outer walls of the castle and the lower part of the cross wall are built with the material laid horizontally, the upper part of the cross wall is built entirely of herringbone work. (p. 63). There is no reason to suppose, therefore, that the causeway at Tamworth is any earlier in date than the keep to which it leads.

The internal buildings at Tamworth date principally from the fifteenth and seventeenth centuries; and the living quarters having been thus kept in line with advancing customs, the castle has enjoyed almost continuous occupation from the time it was built to the present day.

Restormel, Arundel, and Gisors are examples dating from the twelfth century. At Restormel whatever defences the bailey may have had have disappeared; but the keep, though ruinous, retains not only its shell wall to the height of the battlements but also its internal buildings. It is surrounded by a wide and deep moat. (pp. 56, 61). Within the shell there is another

ring wall, concentric with it, and surrounding a large court. The space between the shell and the inner wall, 18 ft. 6 in, wide, is divided by cross-walls into separate apartments, two storeys in height. The ground floor contained the stores and cellars, and the upper floor the halls and living-rooms; the kitchen, near the great hall, rising through both storeys. A square tower, later converted into a chapel, projects on one side of the keep.

The keep of Arundel Castle, which stands on a mound in the middle of a long and relatively narrow bailey, is built of good masonry, faced externally

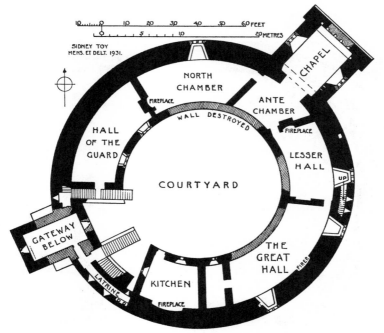

Restormel Castle. Plan of Keep, First Floor. p. 60.

with squared stones and strengthened by pilaster buttresses. Here also a square tower projects from one side of the keep. The original halls and chambers have been cleared away, but they occupied the whole internal space; the roof timbers springing from a central wood pillar and from corbels, which still exist in the shell wall. The central post and some of the cross timbers were taken down within living memory. (p. 53).

The shell keep at Gisors dates from the early years of the twelfth century and from the first had a tower on one side, the foundations of which still remain. The tower was rebuilt in its present form about 1160; but the original structure was not of timber, as has been suggested.[1] Judging from

[1] *Early Norman Castles,* by E. S. Armitage, p. 364.

62   CASTLES

its very substantial foundations it was unquestionably of stone. The shell is multangular in plan, the external angles being strengthened by small pilaster buttresses, and has an internal diameter of about 93 ft. It was approached up the mound by a steep flight of steps which was protected on either side by a wall, long since destroyed. In addition to the main wide entrance there was what appears to have been a sallyport on the west side of the keep. Within the walls on the east side are the ruins, including the apse, of a beautiful little chapel, which was built about 1184 and dedicated to St. Thomas à Becket. On the west side of the keep, near the well, two deep recesses and a large stone wash-basin with a drain through the wall doubtless

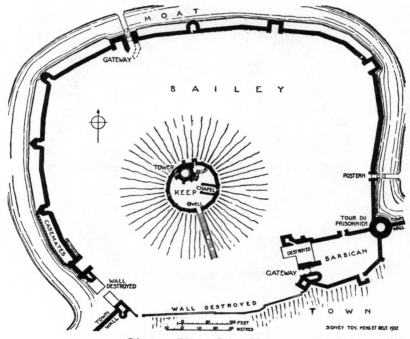

Gisors.   Plan of the Château.

mark the site of the kitchen. The shell wall at Gisors was bonded by heavy timber ties, buried in the masonry. The longitudinal ties were about 12 in. wide by 14 in. deep and 15 ft. long; and there were also cross ties of lesser bulk. The holes left by similar ties were found in walls built at the same period by Bishop Roger at Old Sarum.[1]

The keep at Pfeffengen, ten miles south of Bale, encloses a natural hillock, which rises above the rest of the castle, the shell being built against the

[1] *Proceedings Soc. of Antiquaries*, 1917, 182.

Tamworth Castle. The Causeway. p. 60.

Colchester Castle. Partition Wall in the Keep. p. 71.

Warwick Castle.   The Keep from within the Bailey.   p. 177.

Château d'Arques from the South.   p. 73.

vertical sides and rising up as a mighty tower five storeys above the top of the hillock; it dates probably from the twelfth century. The external face of this great keep is studded all round, from base to parapet, with rough boss-like projections, consisting of stones tailed into the wall and standing out well beyond the face. This form of construction reproduces a method which was employed by the Romans and was recommended by Philo of Byzantium as a protection against attacks by stone-throwing engines. (p. 22).

CHAPTER VII

RECTANGULAR KEEPS OR DONJONS

MEANWHILE a more compact form of stronghold was in process of evolution—the rectangular keep. These keeps provided well-designed and powerful dwelling-houses, which were at once convenient and secure. During the eleventh century they were being built in many places in northern France and, when the initial work of conquest was over, were introduced into England. The White Tower at the Tower of London was begun about 1070, and the keeps at Canterbury and Colchester both about 1080. They were followed in the twelfth century by numerous examples distributed all over England, from Bamborough in Northumberland to Lydford on the borders of Devon and Cornwall.

Generally these keeps are built on the firm ground of the bailey and not on the mound, and stand either completely within the bailey, as at London and Canterbury, or at a strategic point on the curtain wall, as at Kenilworth and Corfe. At Clun Castle, Shropshire, the keep is built on the edge of the mound, its outer walls rising up from the ditch. The small keep at Lydford stands completely upon the mound. (p. 81).

The keeps are strongly built, they have thick walls, generally strengthened by buttresses, and are from two to four storeys in height; each storey often being divided into two or more apartments by partition walls. The entrance doorway is usually on the second storey and is reached by a stairway built against the side of the keep, the stairway often being contained in and protected by a forebuilding. At Arques, Newcastle, and Dover the entrance was on the third storey, the stairway being carried up to this level. From the entrance floor access to the floors above and below was obtained by means of spiral stairways, usually placed in the corners of the keep.

The principal hall was generally on the entrance floor and sometimes had a mural gallery at a high level running round its walls. Mural chambers opened off from the hall and the other apartments. Fireplaces were placed either in the outer or in the partition walls. There was generally a chapel in the keep, built either in the main or in the forebuilding; at Dover there are two chapels, one above the other, and both are in the forebuilding. In some of the keeps, as at Newcastle and Dover, there was a postern or sally-port, providing escape from the keep in the event of its main entrance being carried by the enemy.

The roofs, on account of the combustible and relatively frail materials of

Corfe Castle from the South-East. p. 74.

Tower of London. The White Tower from the South-East. p. 70.

Colchester Castle. The Keep from the South-East. p. 71.

their construction, were vulnerable objects of attack by burning and other missiles. In the early period not only were the roofs constructed of timber but they were often covered with shingles. As a protection for the roofs the walls of the keeps were carried up so high above the gutters as to form a screen behind which the roofs were effectively masked.

Within the keep there was a well, which often went down to a great depth and was carried up through two or three floors, with a drawing place at each floor. Latrines, with one or two turnings in the entrance passage, were formed in the outer walls, opening from the principal rooms.

Among the earliest rectangular keeps in France, now standing at sufficient

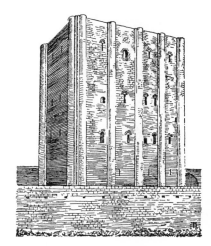

Loches. The Donjon from the South-West. p. 70.

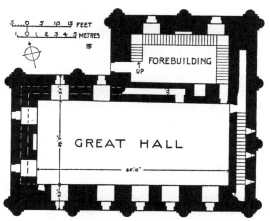

Château de Loches. Plan of the Donjon. p. 70.

height to indicate the original disposition, are those at Langeais, built about the end of the tenth or the beginning of the eleventh century, and Loches, built about 1080.

The keep, or donjon, at Langeais was of much greater length than width. It consisted of a basement and two upper floors; the first floor, apparently, being the store rooms, for there are no windows in the existing walls, and the second floor the living apartments. The entrance doorway is on the first floor. The living-rooms are lighted by round-headed windows set in larger internal recesses, the recesses having straight and not splayed jambs; in the arches of these windows the voussoirs are alternately of tile and stone. At present only two of the walls of this building are standing above the foundations, one of the long and one of the shorter sides. They are faced partly with squared rubble and partly with ashlar, are only 3 ft. 6 in. thick, but are strongly reinforced by buttresses at the corners and along the sides.

The donjon at Loches is one of the most imposing keeps in existence, and stands to-day almost complete. (p. 69). It is built of stone, and consists of a main block and a large forebuilding, the latter in all essentials of its design a part of the main building and originally rising to the same height. The main block measures internally 65 ft. in length by average 27 ft. in width and is therefore more than double as long as it is wide. The walls are 9 ft. 2 in. thick at the base, are strengthened by huge semi-circular buttresses at the angles and on the sides, and rise in four lofty storeys to the height of 122 ft. There are no internal cross walls above the basement. The entrance doorway is on the second storey and is approached by a stairway which rises up on three sides of the forebuilding. The rooms of the main block were lighted on three sides by windows with round-headed internal recesses; the innermost recess being of considerable width and height. The outside openings of the windows are relatively small, and many of them have square lintels. The second floor is reached by a straight mural stairway in the east end wall, and from here a spiral stairway, of later date, leads to the fourth storey and the battlements. The forebuilding, now only three storeys in height, has a chapel on the third storey.

The White Tower at London has a large apsidal projection for a chapel at the south end of the east wall, and a round projection for a circular stairway at the north-east corner. The keep rises in four storeys—a basement and three upper floors—to the height of 90 ft. to the top of the parapet, and from each corner a turret rises high above the parapet. (p. 68). The walls are of great thickness, varying from 12 ft. to 15 ft. at the base; they are built of ragstone rubble with ashlar dressings and are strengthened at the corners and sides by pilaster buttresses. Internally the keep is divided at each stage into two large halls and the chapel, its crypt or sub-crypt.

The entrance was at the second storey on the south; the approach to the doorway being by means of a forebuilding, long since destroyed. The doorway gave access to a large hall, 92 ft. long by 37 ft. wide, which was lighted by windows, now considerably widened, and warmed by a single fireplace. Adjoining is another but smaller hall with a fireplace, and from the smaller hall a doorway leads to the crypt of the chapel. The circular stairway at the north-east corner of the keep, which is 11 ft. in diameter, leads down to the basement, where there is a well, and up to the higher floors and the battlements. The third storey contains the great hall, the withdrawing-room, or solar, and the beautiful vaulted chapel of St. John. There can be little doubt but that originally the great hall and the solar rose to the full height of the keep from this level, as the chapel does; and that the mural gallery which runs round the walls, from the south triforium of the chapel to the north triforium, was open to these rooms. But at a later date the rooms were sub-divided by the insertion of floors at the level of the gallery. From the floor of the third storey two additional corner stairways lead up to the gallery and the battlements.

The keep at Colchester also has an apsidal projection for a chapel at the south-east corner, and in other respects bears some resemblance to the White Tower, though it covers a much larger area. When complete it was probably four storeys in height, but at present only the two lower storeys and fragments of the third remain. The walls are built of rubble with lacing courses of Roman brick and dressings of freestone; they are strengthened at the angles by boldly projecting turrets and at the sides by pilaster buttresses. The main

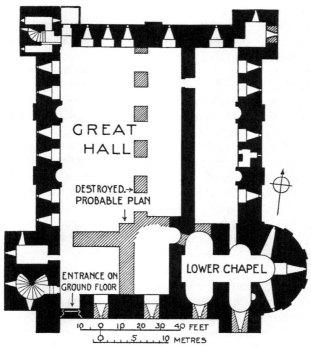

Colchester. Plan of the Keep, First Floor.

entrance, apparently a slightly later insertion, is on the south side, at ground floor level; it was defended by a portcullis and door and was approached by way of a forebuilding, the remains of which were discovered recently. (See also p. 68).

The south side of the keep, where attack was to be expected, was built in an especially solid manner, and contained small rooms including the crypt and sub-crypt of the chapel, and the entrance passage. On the right of the entrance passage there is a well, and on the left a wide spiral stairway to the upper floors. The remaining portion of the floor space, about two-thirds of the whole, was divided by two cross walls, running parallel to each other, into one large and two smaller halls. One of the cross walls has been destroyed, though there are vestiges of it at each end. It was probably solid at ground floor level but pierced by an arcade at the first floor level. The other wall still remains and is of great interest. To the height of the first floor it is built of rubble, with lacing courses of Roman brick, in the same manner as the outer walls; but the upper portion is entirely of herringbone work. (p. 63).

The first floor was probably divided into two halls by the existing cross wall, as at present; the larger hall having an arcade running north and south in line with the destroyed wall; each of the halls was warmed by two large fireplaces. At the north-west corner there is a postern, possibly the original entrance, which had an outer stairway leading to the ground. On this floor also there was a sub-chapel as well as such chambers as existed between the chapel and the west wall. The principal chapel was on the third floor.

Canterbury castle is mentioned in Domesday Book as being then in possession of the King, who had acquired it, or the land on which it stands, from the Abbot of St. Augustine's Abbey. The existing keep must either have been complete or in course of construction about 1086—the date of Domesday. The details, especially of the windows and fireplaces, and the general architectural character of the building clearly indicate that it is to be ranked among the earliest keeps in England. There is no further mention of the keep, as distinct from the castle, until 1173-74, when the sum of £24. 6. 0 was spent on work on the keep and other parts of the castle; and in the following year, when the sum of £5. 11. 7 was spent on the keep alone. These relatively small sums are obviously for repairs or minor alterations.

The keep (p. 73) measures internally 69 ft. by 57 ft. and was originally three storeys in height, but the top storey has been destroyed. The walls are built of rubble with Caen stone dressings, they are 9 ft. 2 in. thick, have a deep battered plinth, and are strengthened by clasping buttresses at the angles and pilaster buttresses at the sides. Internally each storey was divided into four rooms by partition walls, now destroyed.

The principal rooms were on the first floor, with the great hall in the middle and other rooms on either side. The entrance was on this floor; it has been destroyed but appears to have been at the west end of the great hall and was approached by a flight of steps built against the side of the keep. The room at the south-west angle was probably the kitchen. In the corner there is a large circular fireplace of unusual construction. It is 8 ft. 10 in. in diameter and has a large flue, which is carried up through the wall and finished with a domed vault, where the smoke escaped through loopholes in the inner angles of the corner buttress. The lower part of the fireplace, where combustion occurred, is faced with herringbone masonry. A deep recess near by opened on to the castle well, the shaft of which was carried up from the basement to the upper floors and had a drawing place at each stage.

RECTANGULAR KEEPS OR DONJONS

The apartment north of the great hall was probably the retiring room. It has a round-backed fireplace, similar to those at Colchester castle. There is a latrine in this room and another in the room at the south-east corner. The rooms on this floor were well lighted by relatively large windows, having internal jambs which are not splayed but recessed back in three orders, the inside order in one window attaining the width of 15 ft. A spiral stairway at the north-east corner led up to the upper floors and down to the basement. The basement, which was also reached by a stairway down from the kitchen,

Canterbury Castle. The Keep.

was divided into four large store-rooms, two of which were unlighted, while the remaining two received only such light and air as entered by three loopholes set high in the east wall.

The keeps at Arques, Kenilworth, Corfe and Rochester were all built in the first part of the twelfth century, and those at Castle Rising, Newcastle and Dover in the latter part.

The donjon at Arques was built by Henry I about 1125; it has thick walls supported by heavy buttresses and was originally four storeys in height (p. 64). Here the uppermost storey has been destroyed, and the whole structure is in a ruinous condition but the main points of its design are still discernible. The entrance was on the third storey and was approached by a stairway built externally round two sides of the keep and protected by a wall on the outer sides. The stairway was defended from two galleries, which ran along on either side high above the steps and could be occupied by defenders raining missiles on enemies advancing upwards.

When gained the doorway gave access to one room only. For a partition wall, passing longitudinally through the middle of the keep and carried up through three stages, divides each floor into two large halls; and there is no direct communication between them. To pass from one division to the other, or to either of the two floors below, it was necessary to negotiate a complicated system of wall passages and stairways, known only to the initiated. The top floor had no division, and was evidently the quarters of the lord or chief officer. From this apartment the whole keep could be commanded, and here was the only fireplace in the building, and an oven. A well, about 250 ft. deep, was carried up by a circular shaft to the third storey.

The keep at Kenilworth, called Cæsar's Tower, is of exceptionally powerful construction (pp. 75, 78). Although it is only about half the size of the White Tower, its walls are 14 ft. thick and are strengthened by massive towers at the corners and strong buttresses at the sides. It is of two storeys, each storey being of great height. The walls are carried up so high above the level of the gutter as not only to mask the roof but to form two tiers of fighting lines; the battlements and, below the battlements, a row of meurtrières, which are pierced through three sides of the keep, and are approached from a walk at the level of the gutter. Meurtrières are recesses in the wall, sufficiently large for the accommodation of one or two archers, and pierced with loopholes. The keep has no internal divisions but one large hall on each storey.

The entrance, before John of Gaunt's alterations of about 1392, was on the second storey and was approached by a flight of steps built against the west wall. From this level a large spiral stairway at the north-east corner of the keep led down to the ground floor and up to the battlements. At the southeast corner there is a well with a drawing place on each of the two floors. Three of the corner towers were originally solid to the height of the upper floor, and two of them still are, but the lower part of that at the south-west was hollowed out in 1570. All the towers have chambers at the level of the meurtrières and the north-west tower has three tiers of latrines, one for each floor and the third for the battlements. Most of the windows and doorways of the keep have been altered, and the north wall has been destroyed to the level of its base; but many of the original details of the keep are still well preserved.

The keep at Corfe, consisting of two tall stages and a basement, was divided by partition walls into one large and two smaller rooms at each stage.

RECTANGULAR KEEPS OR DONJONS

Here again the outer walls were carried up high above the roof, completely masking it. The lines of the verges of the twin roof are still to be seen on the existing south wall. A blind arcade was carried all round the outside of the keep in line with the roof gutter, but there were no loopholes at this level. Shortly after the keep was built an addition was thrown out across the curtain wall, spanning its walk by a lofty barrel vault (p. 107). One of the most remarkable points of this keep, as indeed of the whole castle, is the

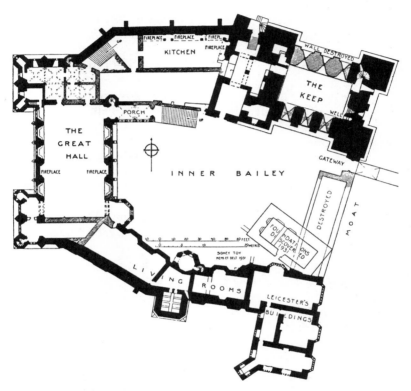

Kenilworth Castle. Plan of the Inner Bailey. p. 74.

excellent quality of its masonry. The walls are composed of a core of chalk and rubble, faced with a very durable limestone, quarried locally, and the mortar used is so powerful that whole masses of masonry, having fallen from as much as 40 ft., when the castle was blown up by gunpowder in 1646, have held together so tenaciously as to remain on the ground unbroken to the present day.[1] (p. 67).

Rochester keep is one of the best preserved, as it is one of the most im-

[1] *Vide* "Corfe Castle" by the Author. Archæologia, Vol. LXXIX, 1929.

posing of its type (p. 77). It consists of a square main building, bisected by a partition wall, and a rectangular wing on one side; the wing containing an entrance porch on the second storey and a chapel on the third storey. The walls of the main building are 12 ft. thick and rise through four storeys to the height of 113 ft.; turrets, above the corner buttresses, rising still 12 ft. higher. The south-east corner was rebuilt in its present form, with a round turret, after the siege of 1215, when a breach was made in the wall.

The entrance was by way of external flights of steps, built against the walls and leading up to the porch. From the porch a doorway, defended by a portcullis, leads into the main body of the keep, and from this level two spiral stairways rise to the upper floors and battlements, and one of them also descends to the ground floor. The great hall is on the third storey; it is well lighted, and the partition here being pierced by an arcade, the hall extends across the whole floor. There are two tiers of windows in the hall, the upper tier with much larger openings than the lower one, being at the level of a mural gallery which runs round the walls 14 ft. above the floor level. There are two fireplaces, one in each of the north and south walls. The arches of the arcade, as well as those of the internal window recesses, and of the fireplaces, are all enriched with mouldings and chevron ornament. A well shaft with a drawing place at each floor is carried up through the partition wall of the keep.

In each of the keeps at Castle Rising, Newcastle, and Dover the approach stairway is enclosed within and defended by the forebuilding. At Castle Rising, as at Rochester, the entrance is through a porch at second storey level; here, however, the keep is only two storeys in height. A noticeable feature of this keep is the extent of its external ornament, unusually great for a military building. But the ornament is so disposed, especially in respect to the long vertical mouldings of the buttresses, as rather to increase than detract from the impression of strength and dignity.

The keeps at Newcastle and Dover have many points of resemblance to each other in design. Both have their entrance doorways at third storey level, and in each case the long approach stairway is enclosed, and strongly defended all the way up, by the forebuilding. Both also have many mural chambers constructed in the thickness of their walls, and a mural gallery, high above the floor, running round the walls of their great halls.

Newcastle keep, which is much the smaller building, has no internal partition, each floor consisting of the central hall and the surrounding mural chambers; it was built 1172–1177. It consists of a vaulted ground floor and two upper storeys; the walls rising sufficiently high above the roof as to mask it completely. The chapel, consisting of chancel and nave, vaulted and richly decorated with chevron and other ornament, occupied the whole of the ground floor of the forebuilding; the stairway to the keep passing over its roof. It is entered at ground floor level from outside, and there was no

Rochester Castle.   The Keep from the North-West.   p. 75.

Launceston Castle. The South Gateway from without. p. 105.

Kenilworth Castle. The Keep from the South. p. 74.

doorway between the chapel and the other parts of the keep; indicating that the chapel was for the use of the garrison in general. The great hall, on the third storey, is very lofty; and its mural gallery, with openings to the hall at the ends of the original ridge roof, is 30 ft. above the floor; the existing vault is modern. From the hall a spiral stairway at the south-east corner of the tower descends to the basement and rises to the battlements, while a straight flight of steps leading out of the stairway at the hall level passes up through the east wall and ends in a second spiral stairway at the north-east corner, which also rises to the battlements. From the first stairway there is a postern which opens out on the face of the forebuilding, high above the ground, and originally led by a gallery and bridge to the curtain wall of the castle, so that in the event of the keep being taken the defenders on the battlements could escape by either stairway and gain the postern without passing through the great hall.

The great square keep at Dover, built 1181–1187, consists of a basement and two upper storeys and rises with two slight off-sets to the height of 83 ft. to the top of the wall; square turrets at the corners rising still 12 ft. higher. The walls are well buttressed and of great strength, varying in thickness from 17 ft. to 21 ft., and contain many mural chambers at each stage. The forebuilding is strengthened by three towers and contains three long flights of steps which begin at the south side of the keep and continue up the east side. At the head of the first flight of steps there is a chapel, richly decorated with clustered pillars and chevron moulded arches, and at the head of the top flight there is a guard-room, facing down the steps and defending the main entrance on its right. (p. 80).

The main building is bisected by a cross wall which ascends through the full height of the keep, dividing each floor into two long halls. The third storey or entrance floor contained the principal rooms, consisting of two large halls and the surrounding mural chambers; the upper gallery is carried all round the outer walls of both the large halls. From one of the mural chambers a narrow passage leads to a second chapel, which is constructed in the forebuilding immediately above the lower chapel and must have been for the use of the occupants of the keep. The existing brick vaults over the halls were built in 1800 for the support of artillery on the flat roof.

A mural chamber on the left of the entrance doorway contains the well. Harold's well, as it is called, is about 250 ft. deep and is lined with masonry to the depth of 172 ft below the mouth. A recess beside the well contained a tank from which supplies of water were conveyed, through lead pipes buried in the thickness of the walls, to other parts of the keep. Two of these lines of pipes have been traced and there are indications that there are others not yet definitely followed. A system of water supply, similar to this at Dover, also exists at Newcastle, where pipes lead from a recess in the well chamber to other rooms in the keep.

From the entrance floor two large circular stairways, one at the north-east and the other at the south-west corners of the keep communicated with the lower floors, the mural gallery and the battlements. From a mural chamber on the second storey there is a postern opening on to the forebuilding above the stairway. This doorway was rebuilt in Tudor times, but probably from

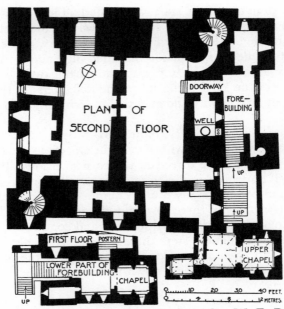

Dover Castle. Plan of the Keep. After W. E. Peck.

the first it was a sally-port, to be used either for escape or for a rear attack on an enemy advancing up the stairway.

On the Ile St. Honorat, off Cannes, there is a keep which stands alone without any outside defences and appears to have been in course of construction between 1073 and 1190. It was built by the monks of the Abbaye de Lérins as a retreat in time of trouble and was restored in the fourteenth century and again in the fifteenth century, when the machicolated parapet was added. It is an irregular shaped tower-like structure several storeys in height.

Lydford Castle.   The Keep from the South.   p. 66.

Chepstow Castle.   Fitz Osbern's Hall from the South.   p. 101.

Roumeli Hissar from the West, looking across the Bosporus to Anadoli Hissar on the Asiatic shore. p. 83.

CHAPTER VIII

BYZANTINE AND SARACEN FORTIFICATIONS OF THE TWELFTH CENTURY

FORTIFICATIONS in the Levant during the eleventh and twelfth centuries continued to follow the traditions of the Byzantine empire. The towers were usually circular and the north or Black tower of the castle of Roumeli Hissar, built probably by Alexios Comnenus about 1100, may be taken as an example. The fortifications of Roumeli Hissar, the castle of Europe, and Anadoli Hissar, the castle of Asia, stand facing each other across the Bosporus at the narrowest point of the Straits, about seven miles north of Constantinople. They were built by the Greeks to protect the city against raids from Russia and other countries round the Black Sea, but were remodelled and considerably enlarged after the Turkish Conquest.[1] (p. 82).

The structure, called the Black Tower on account of its grim associations, is 81 ft. in diameter, its walls are 24 ft. thick, and it rises in seven storeys to the height of 120 ft. The two upper storeys were added after the Turkish conquest in 1452; but originally, as at present, there was a broad fighting platform behind the battlements, and the roof was screened by a high encircling wall. The tower is faced throughout with coursed stonework, and at the base has a deep battered plinth. (pp. 84, 87).

The entrance doorway is on the ground floor and from it a passage led straight through the wall to the central chamber of that floor. Opening from the right of the passage there is a spiral stairway, which ascends to the level of the original battlements, and on the left is a circular mural chamber with a domed roof. Round the central chamber are three wide recesses and two doors, each door leading to a domed chamber similar to that from the passage. The second storey has neither recesses nor mural chambers, but on each of the third, fourth, and fifth storeys there are two large recesses; a mural chamber near the stairway and one or two latrines. The ancient timber flooring still exists in all the stages, though its condition is now decayed and treacherous. A dark passage near the head of the stairway leads to the crown of a deep circular oubliette, which is constructed in the thickness of the wall and has no window or any other entrance than this passage. There is no door or barrier at the point where the passage enters the oubliette, and a prisoner impelled along the passage and pushed through the opening would fall in utter darkness to the bottom of the chamber 13 ft. below. This is probably one of the earliest examples of a true oubliette, of which there are very few.

[1] *Vide* "The Castles of the Bosporus," by the Author. *Archæologia*, Vol. LXXX, 1930.

M. Viollet-le-Duc in the course of his extensive investigations found only two; one at Pierrefords and the other at the Bastille, Paris. What are actually reservoirs or latrines have often been called by this name. An oubliette was a

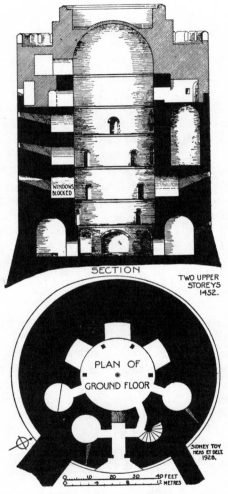

Roumeli Hissar. The Black Tower.

secret prison, having its only opening at or near the roof and into which prisoners destined to "disappear" were thrown, or into which they unwittingly fell.

During the Byzantine period the Black Tower was probably at once a fortress and a prison. The upper rooms, being for the soldiers, were amply provided with fireplaces, even the recesses, which were probably the sleeping quarters, having a fireplace each. The ground floor with its recesses and mural chambers, where there are neither latrines nor fireplaces, were for the accommodation of the prisoners. Even in respect to the soldiers quarters it would be difficult to imagine any place more terrible than this tower. From floor to roof the tower was in all but total darkness, and with the exception of one mural chamber on the fourth floor, received only such light as filtered into it through narrow loopholes. At some period, probably after the Turkish conquest when the whole tower was used as a prison, even the loopholes were blocked, as they still are. It is little wonder that on account of its gloom, its strength, and its painful associations, this building was regarded with extreme aversion, and that at the Byzantine period it was known as one of the towers of Lethe, or oblivion.

The entrance doorway of the Black Tower is protected by a machicolation, which passes 25 ft. up through the wall and was commanded from the fourth storey.

Machicolations were holes formed in the roofs of gateways and entrance passages through which boiling pitch, stones, darts and other missiles were thrown on the heads of assailants below. When over the entrance to the gateway they also enabled the defenders to quench fires lighted by besiegers to burn down the gates; and this appears to have been their original purpose. Flavius Vegetius, writing about 390, says: "It is necessary also to have a projection above the gate with openings from which one can pour water on the fire which the enemy has lighted."[1] Machicolations were also built on the crests of walls and towers to repel the operations of sappers at the base. In this position they first took the form of hoards or brattices, timber platforms projected out from the battlements in times of siege, which, in one form or another, had been in use from about 1500 B.C. By the end of the twelfth century A.D. temporary hoarding gave place to machicolations in stone in many new buildings; but it was not until the end of the thirteenth century that this custom became general, and hoarding was employed at a much later date in older castles. As constructed in stone the parapets were built out on corbels about a foot beyond the outer face of the wall; the corbels being spaced sufficiently far apart as to allow of a large hole, or machicolation, between each pair of corbels.

Among the finest military works of this period are the walls and towers at the north end of the land walls of Constantinople, between the towers of Anemas and the palace of Porphyrogenitus, built as a protection to the palace of Blachernæ. A large portion of this wall, built by Manuel Comnenus, 1143–1180, is constructed with internal arcades of lofty arches, and is 15 ft. thick at the top. It is strengthened at frequent intervals by massive towers of

[1] *Vegetius*, Book IV, C. 4.

varying shapes, round, octagonal, and square, and all projecting entirely on the outside of the curtain. Both wall and towers are built of large blocks of stone with brick lacing courses, several bricks thick. A powerful tower, also of the twelfth century, immediately north of this portion of the wall, was designed to be at once a residence and a fort.

The castles at Roumeli Kavak and Anadoli Kavak, guarding the Bosporus near its northern end, were built about the middle of the twelfth century. They stand facing each other on opposite shores, and from each of them a wall ran down the hill to the water's edge. The castle at Roumeli Kavak has

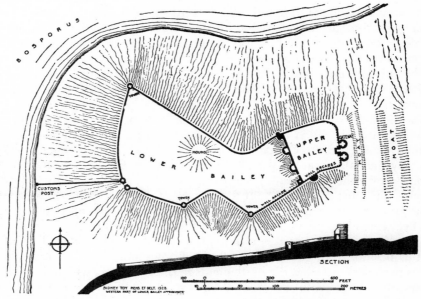

Anadoli Kavak. Plan of Hieron Castle.

been destroyed, but Hieron Castle at Anadoli Kavak on the Asiatic shore, though ruinous, still retains its walls and towers to a considerable height.

Hieron Castle consists of an upper bailey, on the crest of the hill, and an extensive irregular shaped lower bailey, which runs down the hillsides towards the Straits. (See also p. 124.) It is built of large coursed stonework with brick lacing courses, seven bricks thick, and bears marked resemblance in its masonry to that portion of the land walls of Constantinople built by Manuel Comnenus. The walls of the lower bailey are strengthened at rare intervals by towers, circular on the outside and cruciform within, each tower having a lofty lower stage covered by a cupola vault. A portion of the south wall is constructed with internal arcades.

The upper bailey is roughly quadrangular, and the south and east walls,

Roumeli Hissar. The Black Tower from within the Castle. p. 83.

Skenfrith Castle. The Keep from the South. p. 99.

Longtown Castle. The Keep from the West. p. 98.

BYZANTINE AND SARACEN FORTIFICATIONS

those most exposed to attack, are constructed with internal arcades. The wall between the baileys is strengthened by four strong towers, and the only other wall tower is in the south wall. The east side of the bailey, where the ground rises slightly, is the most vulnerable side of the castle, and here there is a powerful gatehouse, flanked on either side by a strong round tower and defended by two lines of moats.

With the powerful Byzantine fortifications which the result of conquest placed in their possession, the Saracens were able to defend themselves for long periods against the soldiers of the first Crusade, 1096–1099. Behind the walls of Nicæa, Antioch, and Jerusalem, they offered a stubborn resistance to the might and skill of Western Europe. Again and again the enemy was repelled discomfited; and it is no small tribute to the skill, resource, and intrepidity of the Crusaders that they were successful in that expedition.

The Saracens built their own military works largely on Byzantine models but, in doing so, profited by the great experience they had gained in defence and attack. Their work reached its zenith under the powerful Saladin, Sultan of Egypt, who between 1170 and 1182 built the walls and citadel of Cairo. These fortifications are among the finest works of military architecture built during the whole of the Middle Ages. Scientific design and constructive skill are particularly marked in the gateways and towers.

The walls of the citadel of Cairo, as left by Saladin, are 9 ft. 2 in. thick, are built of good masonry, and rise to a great height above the sand. They are protected at the corners and at frequent intervals along the sides by round towers, each containing two tiers of chambers with meurtrières. The mural galleries are lighted by windows on the inside face of the wall. Round the whole of this fortress there were three tiers of fighting lines; two from the towers and mural galleries, and the third from the battlements.

Many additions and alterations were made to the citadel after the time of Saladin, particularly during the reign of his immediate successor, who added large square towers at various points in the wall, and in the sixteenth century, when a projecting portion of the fortress at the south-west was cut off by a cross wall having a great circular tower at each end and a gateway in the middle. But the main portion of the curtain and the wall towers are the works of the great Sultan, as are three of the gateways. Each of these gateways is built on an L-shaped plan; the inner arch being placed at right angles to the outer gate, involving a right-angled turn to all those passing through the entrance.

CHAPTER IX

TRANSITIONAL KEEPS OF THE TWELFTH CENTURY

DURING the whole of the twelfth century there was a constant stream of military forces passing to and fro between Europe and the Levant. The war between the Crusaders and Saracens, far from being confined to the main crusades, was incessant, and the expeditions to Palestine continuous. On their return from these expeditions the Crusaders proceeded to apply to their own fortifications the knowledge and experience they had gained in their recent campaigns abroad.

The rectangular keeps and towers built at that time in Western Europe had the great disadvantage of presenting vulnerable corners to the sapper and the battering-ram, since the enemy could be attacked from one side only, and was sheltered by the corner itself against attacks from the other side. In the Levant the towers were frequently circular or multangular, and therefore presented no screen to the enemy at any point. On the other hand a rectangular plan is much more convenient for the disposition of the rooms inside a building than a circular plan, and the development of the latter form from the former was very gradual in Western Europe. It was not until the return from the third Crusade at the end of the twelfth century that round keeps were built generally.

Meanwhile many keeps of a transitional character, combining the advantages of both forms, were built, and conspicuous among them are those at Houdan, Seine et Oise; Provins, Seine et Marne; Gisors, Eure; Orford, Suffolk; Etampes, Seine et Oise; Conisborough, Yorks.; and Longtown, Herefordshire.

The donjon at Houdan was built about 1130. On the inside the plan is square, with splayed corners; but on the outside it is circular, with four projecting turrets, each turret covering one of the internal corners (p. 91). There are two storeys, each of great height. The ground floor was the storeroom and contains a well. On the upper floor was the great hall with chambers, formed in three of the turrets, opening off from the hall. The fourth turret contains a spiral stairway leading from the hall to the battlements. The walls are carried up on all sides 16 ft. above the level of the gutter to mask the roof.

The present entrance to the donjon, broken through the wall at ground level, is modern. The original entrance doorway is in the north turret, 20 ft. up from the ground, and was reached by a drawbridge from the curtain wall

TRANSITIONAL KEEPS OF THE TWELFTH CENTURY 91

which passed near the donjon at this point. The doorway gained, a straight mural stairway on the left leads up to the hall on the first floor; and from one of the chambers opening out of the hall a short mural passage and a spiral stairway lead down to the ground floor.

Keeps of the middle and latter part of the twelfth century sometimes

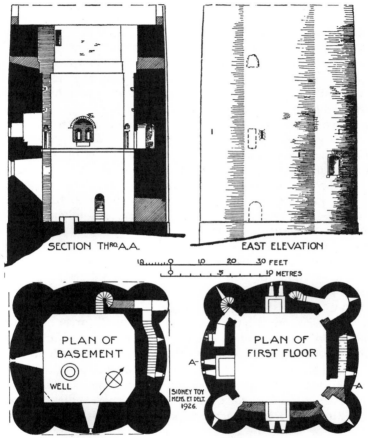

Château de Houdan. The Donjon.

occupied strategic positions on the curtain wall of the bailey, or on a shell wall; as at Conisborough and Gisors. When built detached within the bailey they were often surrounded by a narrow court and a high wall called a chemise. The approach to the keep was then by a gateway through the chemise to the court, and by a flight of steps within the court up to the wall walk of the chemise. At one point a causeway or wall, with a drawbridge at the end, was built across the court between the wall walk of the chemise and the entrance doorway of the keep, and it was across this narrow causeway, exposed to attack from all sides, that an enemy had to defile, if he wished to gain the keep. The Tour de César, Provins, and the donjons at La Roche Guyon, Etampes, and Chatillon-sur-Indre, were all built on this design.

The Tour de César at Provins was built about the middle of the twelfth century, and remains to-day one of the most interesting and best preserved of the transitional keeps. It stands upon a mound and is surrounded by a battlemented wall, or chemise, which at one point extends down the mound into the bailey, enclosing a vaulted stairway of approach. Internally the plan of the tower is similar to that of the donjon at Houdan. But externally it is square at the base and octagonal above, with four semi-circular turrets rising from the corners of the base. (p. 93).

The entrance to the tower was first by way of the vaulted stairway up the mound, already mentioned, to the space between the chemise and the tower. From here a stairway led to the battlements of the chemise, and thence to a causeway and drawbridge across the interval between the chemise and the entrance doorway of the tower. From this doorway a short flight of steps, guarded on one side, led up to the great hall on the upper storey. In 1432 the upper part of the chemise was taken down and the causeway to the tower replaced by steps up from the court.

The tower is of two storeys, both covered with stone vaults, domical on the underside and flat on the upper surface. The lower storey, or basement, consists of a large room, lighted by narrow windows placed high in the wall, and a mural chamber, which was probably a prison. From one side of the large room a mural passage and steep flight of steps led down to a well.

The upper storey contains the great hall and small turret chambers. Here the chambers do not open directly out of the hall, as at Houdan, but either from mural passages or stairways communicating with the hall. A doorway in the middle of each of the north, east, and west walls leads straight through to a small outside platform. M. Viollet-le-Duc suggested that from these platforms three other drawbridges and causeways radiated towards the chemise.[1] But while such an arrangement, allowing the enemy, on gaining the chemise, four points of attack on the tower instead of one, would be totally contrary to the principles of mediæval defence; the shape of the platforms, incorrectly drawn in the *Dictionnaire*, and the remains of the enclosing walls indicate that these structures were small defensive turrets. As seen from the toothing on either side, the small bracket-shaped platforms were each enclosed by a wall, with, probably, three loopholes, one on the front and one on each flank. The fireplaces on this floor and the basement are of later date.

[1] *Dictionnaire Raisonné de l'Architecturé.* Tome V, p. 64.

TRANSITIONAL KEEPS OF THE TWELFTH CENTURY 93

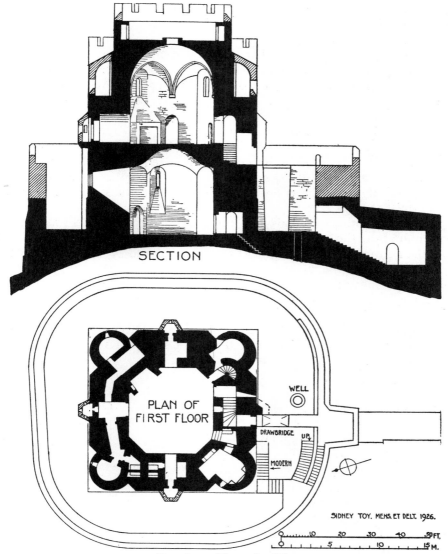

Provins. Tour de César.

From the hall one stairway leads down to the basement and another up to an external gallery, which runs round the walls of the upper part of the hall and had loopholes, now destroyed, to the field. And, finally, narrow mural stairways constructed round the back of the dome and opening midway to the battlements of the turrets, lead from the gallery to the upper battlements.

Since the gallery occupies the usual position of windows, the hall is lighted by a large circular eye in the summit of the dome, 35 ft. above the floor. The present battlements and pointed timber roof were built in the seventeenth century, and the original form of roof is a question of conjecture only. But it is very probable that the eye was open to the free passage of light, as in the donjon at Coucy and the Tour de Constance at Aigues Mortes; the screen walls, which rose high above the upper face of the vault, being adequate protection against enemy missiles.

At Gisors the donjon built by Henry II about 1160 replaced and stands partly upon the foundations of an older tower, built with the shell wall some fifty years before. It forms an irregular octagon in plan,[1] is strengthened at the angles by massive buttresses and is four storeys in height. Forming as it does a part only of the larger enclosure within the shell it is less than half the size of the Tour de César.

The original entrance stands only 8 ft. above the ground, and was reached by steps up from the court. Near it is a sally port, through the shell wall; and both the entrance and the sally port are obscured and protected behind one of the buttresses of the donjon. (p. 62).

From the entrance doorway a straight stairway in the wall led up to the first floor, and from here a spiral stairway rose to the upper floors and the battlements. The ground floor, now broken into by a doorway, originally had neither window nor ventilation, and must have been reached through a trap-door in the first floor. Each of the upper storeys has one window, and in the third storey, there are two latrines. There is no fireplace on any floor. Towards the end of the fourteenth century a separate spiral stairway, with an outside entrance and a doorway to each of the upper floors, was built against the east side of the tower and carried up as a turret above its battlements. The original entrance and mural stairway were then blocked.

Orford keep, also built by Henry II, was begun in 1166 and finished in 1170; it is a much more substantial building than the donjon at Gisors. The plan is circular internally, but externally it is multangular, with three square turrets equally spaced round the keep and projecting out boldly from its face. A forebuilding containing the entrance porch, with a basement below and a chapel above, is built in the angle between the south turret and the main body of the keep.

The keep consists of a basement and two upper floors. The walls are carried up high above the gutter level to mask the roof, and the turrets rise about 20 ft. above the battlements. In the basement there is a deep well, which is lined with dressed stonework and, cut in the stonework, has handholds and foot-holds for descent to the bottom.

The entrance into the keep is by a straight flight of steps to the porch of the

[1] Enlart: *Manuel d'Archéologie Française,* Vol. II, 505, and Armitage: op. cit. 364, incorrectly describe it a decagonal.

TRANSITIONAL KEEPS OF THE TWELFTH CENTURY 95

forebuilding, and from the porch by a passage, barred by two doors, to the large central hall on the first floor. Both this hall and that on the floor above are well lighted, and each of them has a fireplace.

One of the turrets contains a spiral stairway, running from the basement to the battlements. The others contain chambers, and since these chambers if

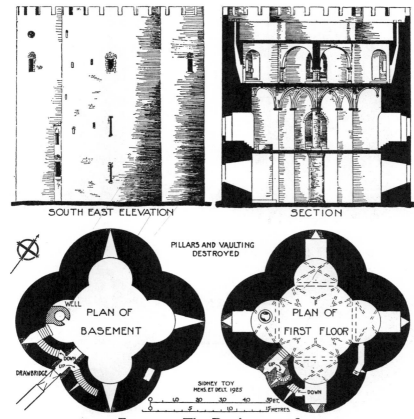

Etampes. The Donjon. p. 96.

made the same height as the halls would be disproportionately lofty, they are subdivided into two stages for each of the halls. The lower chambers are entered by mural passages opening from the window jambs in the halls, and the upper chambers by mural passages leading round the walls from the stairway. One of the chambers of the first floor, with two large fireplaces and a lavatory or sink, is the kitchen. The turrets have two other tiers of chambers above the level of the roof gutter, one of the chambers containing a double oven.

96　　　　　　　　　　　CASTLES

The donjon at Etampes, dating probably about 1160, represents another form of transition from square to round keeps, in that while retaining spacious halls and living apartments it assumes a rounded form. It is a large quatrefoil-shaped building, three storeys in height; its floors and roof originally being supported on a central pier, long since destroyed. The

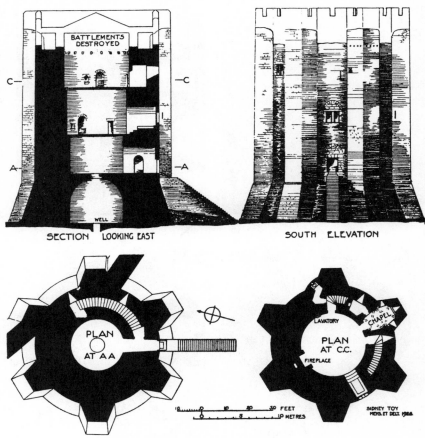

Conisborough Castle. The Keep. p. 97.

ground floor, where there is a well, had a timber ceiling, but the first floor, forming the great hall, was covered with a very fine and complicated rib vault, vestiges of which still remain. The walls are carried high above the upper storey, as a screen for the roof. (p. 95).

The entrance doorway is placed midway in height between the ground and first floors, and was reached, probably, by means of a wall, or causeway, projected out from the chemise towards the donjon, and having a drawbridge

TRANSITIONAL KEEPS OF THE TWELFTH CENTURY

at the end of it. The doorway opens to a passage which has mural stairways on the right and left, that on the left leading down to the store rooms on the ground floor, and that on the right up to the great hall. The entrance passage goes straight through the wall; so that an enemy, who had forced the entrance and, unaware of trap, had rushed straight ahead, would fall on the pavement of the ground floor, 12 ft. below. From the hall another mural stairway led to the third storey and from there a spiral stairway rose to the battlements.

Conisborough, one of the finest keeps in England, was built about 1180–90. It is a tall cylindrical tower with very thick walls supported by six massive buttresses; the buttresses rising to the full height of the building. It consists of a vaulted ground storey, or basement, and three upper floors. Neither the basement nor the first floor has any windows, and the only access to the

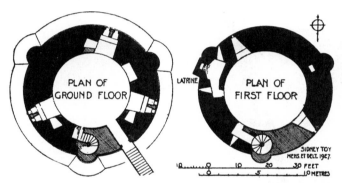

Longtown Castle. The Keep. p. 98.

former, which contains a well, is through a large hole in the centre of the vault (p. 96).

The entrance doorway is on the first floor and is approached now by a modern flight of steps up from the courtyard; originally there was, probably, a similar flight of steps and a drawbridge before the doorway. From the first floor a mural stairway, built concentrically with the wall, leads up to the second floor, and similar stairways rise to the third floor and the battlements. Each of the second and third floors has a two-light window, a large fireplace and a stone lavatory basin, and there is also a latrine on each floor. A vaulted chapel with a small sacristy, built partly in the wall and partly in one of the buttresses, opens out on the east side of the third floor.

The battlements have been in a large measure destroyed. But it is clear from the parts which remain that the wall and buttresses were carried up to form a screen round the roof, and that there were two fighting lines; one from a gallery which runs round the roof at the level of the gutter, and the other from the battlements, about 12 ft. above the gallery. Three small vaulted chambers and an oven, all formed in the buttresses opened on to the gallery. The oven, like those at Orford, was probably for the use of the fighting forces, though it may have served a domestic purpose also. It is 5 ft. 8 in. in diameter, 3 ft. 7 in. high to the summit of its domical roof, and has a rectangular depression in its floor 16 in. by 19 ins. and 6 in. deep. One

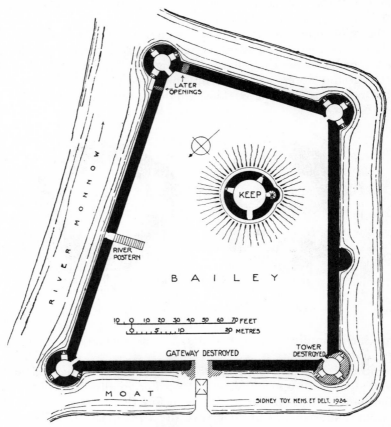

Plan of Skenfrith Castle. pp. 99, 100.

of the vaulted chambers is pierced by numerous pigeon holes, about 6 in. square. It is probable that the gallery was covered with a roof or vault supporting the wall walk above, for it had an inside wall 3 ft. 8 in. thick. Two stairways from the gallery led up to the battlements.

The keep at Longtown, built about 1180, is also cylindrical, but here the circular wall is supported by three half-round buttresses, one of which provides the necessary thickness for a spiral stairway (pp. 88, 97). The

TRANSITIONAL KEEPS OF THE TWELFTH CENTURY

keep stands upon a high mound; it is of two storeys, and has a battered plinth 12 ft. high. The entrance doorway and the walling above it have been destroyed. But the doorway appears to have been in the lower storey, the floor of which was about level with the top of the plinth, and to have been approached by a flight of steps up from the foot of the mound. Here also the walls were carried up sufficiently high to mask the roof.

Differing from most of the other keeps the principal room at Longtown was on the ground floor. This room had three windows, each of two lights, and a wide fireplace. In one jamb of each of the windows there is a large cupboard recess. The upper room was relatively low, was lighted by loopholes only and has no fireplace; a doorway on one side opens to a latrine. The spiral stairway rises from the ground floor to the battlements. Reichenberg, Bohemia, has a keep similar in plan to that at Longtown.

Skenfrith Castle, Monmouthshire, about 1190–1200, has a cylindrical keep with one semi-circular projection on its face (pp. 88, 98). Here the projection is exclusively for the accommodation of a spiral stairway, which was too large to be constructed within the wall itself. The upper part of the building has been destroyed; at present the keep consists of a basement and two upper floors. The basement was lighted by narrow loopholes placed high in the wall, and must have been entered by a trap-door in the floor above. The entrance doorway was on the first floor and the spiral stairway ascended from this level. Excavations made in 1925 showed that the mound is artificial, composed of rubble and sand. The foundations of the keep, with rough faces on both sides, are carried down through this material to the natural soil.

CHAPTER X

FORTIFICATIONS AND BUILDINGS OF THE BAILEY IN THE ELEVENTH AND TWELFTH CENTURIES

THE curtain walls surrounding the baileys of eleventh and early twelfth century castles were sometimes perfectly plain, and had no other defence than their battlements, as at Eynesford in Kent and Trematon in Cornwall. But in many cases they were strengthened at strategic points by square towers, projecting on the outside. Where they occurred, the wall towers were either spaced widely apart, as at Ludlow, or concentrated on the more vulnerable side of the castle, as at Richmond. Later on in the twelfth century the towers were built closer together, in order to command more effectively the panels of wall between them, as at Dover. At Gisors two of the wall towers on the west side of the castle, about 1180, are built with spurs, like the cutwaters of a bridge, the spurs being carried to the full height of the towers (p. 62).

When the castle stands upon high ground the curtain wall follows the irregular contours of the site, as at Richmond and Ludlow. But when upon level ground the walls are built with long straight sides, as at Sherborne and Skenfrith. The curtain wall at Skenfrith is composed of four perfectly straight sides with a tower at each corner. Each of the corner towers consists of a basement and one upper storey, and it is worthy of note that the floors of the upper storey in all three remaining towers—one has been destroyed—were all on exactly the same level, as though they had been set out by some delicate and precise instrument (p. 98).

On level sites the curtain wall was surrounded by a moat, as at Sherborne; or defended partly by a river and partly by a moat, as at Skenfrith. But when standing on a hill the castle is generally defended on one or more sides by precipitous rocks or steep declivities. The moat was then confined to the side of the castle where the approach was more gradual and the entrance gateway was usually placed on that side.

HALLS

Generally in the more important castles the buildings within the bailey included a large hall for the common life of the garrison. The hall sometimes stood near the middle of the bailey, but more often it was built against the curtain wall; the curtain itself forming one of its sides. A kitchen generally stood near the hall, if not adjacent to it; and there was often a well near

FORTIFICATIONS AND BUILDINGS OF THE BAILEY

the kitchen. In some early examples the hall was virtually the keep of the castle, and such was the great oblong hall built by William Fitz Osbern at Chepstow about 1070.

The hall at Chepstow Castle, now called the keep, stands on a narrow tongue of rock between the river Wye and a deep ravine (p. 81). It is a powerfully built structure consisting of a basement and two upper storeys and measuring internally 89 ft. long by average 30 ft. wide; the walls on the east and south, the most vulnerable sides, are over 8 ft. thick. Though the upper stages were remodelled in the thirteenth century by the insertion of new windows and a transverse arch, the walling is original and retains the original windows in the basement and internal wall arcading on the south and west sides of the first floor.

At Richmond Castle, Yorks., there is a long rectangular hall in one corner of the bailey which is contemporary with the earliest part of the curtain wall, about 1070-80. Here the hall with the living rooms and domestic offices adjoining formed the living quarters of the ruler of the castle. The existing keep was not built until the following century.

At the castles of Sherborne and Devizes, both built about 1120 by Roger, Bishop of Salisbury, the hall and its domestic offices were built within the bailey, near the keep and well away from the curtain walls. The hall at Sherborne is still standing, though ruinous; it is part of a group of buildings forming a complete castle within a castle (p. 102). This group stands in the middle of the bailey and is built round a square courtyard. The hall is in the north range, the domestic offices on the east of the courtyard, and other buildings, now destroyed, on the south and west. The great rectangular keep occupies the south-west corner of the group. The hall at Devizes has been destroyed to the foundations. It was a large building with arcades on either side, dividing the hall into a central nave and two side aisles.

The hall of Leicester Castle, built about 1150, also has a nave and two aisles, though here, while the walls are of stone, the pillars and struts between the nave and the aisles are of oak. It stands within the bailey with its axis running north and south, and although considerably altered its original disposition is still clear. On the south of the hall were the kitchens, destroyed in 1715, and beyond the kitchens there is a vaulted undercroft measuring internally 50 ft. by 18 ft. The living quarters, now completely destroyed, were doubtless on the north side of the hall on the site of what was called the Castle House. Near the hall there is a mound, about 30 ft. high, which probably formed at one period the stronghold of the castle, but since the mound was lowered to the extent of about 15 ft. a hundred years ago it is not possible to say now what buildings stood upon it.

One of the best preserved halls of the twelfth century is that at Oakham, Rutland. This hall consists of a nave and two aisles, all of stone and having arcades richly decorated with leaf and dog-tooth ornament. It was built, together with kitchens and living rooms, long ago destroyed, about 1180. Oakham, however, was not a castle but a manor house, defended by a curtain wall and ditch.

At Durham Castle a chapel was built against the curtain at the foot of the mound in the eleventh century, and in the latter part of the twelfth century a long hall, called Constables Hall, was added to the west of the chapel.

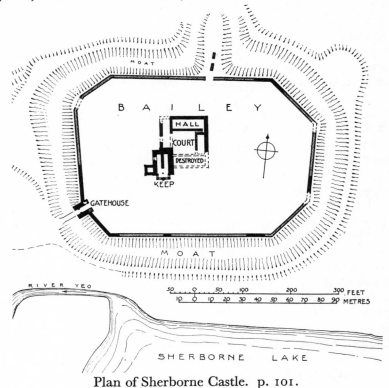

Plan of Sherborne Castle. p. 101.

CHAPELS

The chapel held an important position in the life of the castle. When not situated in the keep, as described above, it generally stood in the bailey, as at Durham and Gisors. At Richmond, Yorks., the chapel, dating from the eleventh century, is in one of the wall towers; and the chapels at Castle Rising, Bamborough, and Ludlow, all built in the twelfth century, stood isolated in the courtyard.

Those at Castle Rising and Bamborough have been destroyed, but the circular nave of the chapel at Ludlow, with richly-carved arches and wall arcade, still remains.

FORTIFICATIONS AND BUILDINGS OF THE BAILEY

GATEWAYS

The gateway into the bailey was generally a substantial building of two or three storeys, and was approached from across the moat by means of a causeway and a drawbridge. In addition to the main gateway there was usually a small postern, placed in such a position that escape could be effected or a sally made unobserved by the enemy. At Dover there are two large gateways in the curtain, one at the east and the other at the west of the bailey, and each of them was defended by an outwork or barbican.

Of gateways dating from the eleventh century that at Exeter Castle was built about 1070, and that at Ludlow, about 1090. The former projects entirely outside the curtain, while the latter, with only a slight projection on the outside, extended about 40 ft. within the bailey. In each case the gateway is surmounted by a tall strong tower, which was virtually the keep of the castle.

The gateway at Exeter Castle consists of the main body of the building, three storeys in height, and a lofty barbican, which projects a short distance outside the entrance arch (p. 104). The passage through is spanned at either end by a round arch of two orders, 10 ft. 3 in. wide. The outer, or entrance, arch is now blocked and its details obscured by masonry, but it was probably closed by a two-leaved door, secured by a timber bolt. A moat passed along the curtain in front of the gateway and was probably spanned at the entrance by a drawbridge. The barbican is formed by two deep buttresses, which project out in line with the lateral walls of the gateway, and an end wall, built on a lofty arch that spans the space between the buttresses at the height of the second storey. Its end wall is therefore completely open up to a height sufficient to admit light to the windows above the gateway. The barbican was entered from the second floor of the gate tower and formed a powerful fighting platform, commanding not only the field and both flanks but also the entrance to the gate below. The upper storeys of the tower must have been approached from the curtain wall, for there is no stairway from the entrance passage.

The gatehouse at Ludlow Castle, called the Great Tower, is one storey higher and much more substantially built than that at Exeter. About a hundred years after it was built a new entrance was made through the curtain near the tower and this gateway passage was blocked at both ends, covered with a stone vault, and converted into a prison. The cross arches of the gateway have been destroyed and the north wall of the tower has been rebuilt; but as the result of investigations made in 1903-4 the original disposition and plan of the gateway passage have been recovered.[1]

The passage was divided by a cross arch and doorway into a short entrance porch 8 ft. 6 in. long, and a gate-hall, 29 ft. 6 in. long; the walls

[1] "The Castle of Ludlow," W. H. St. John Hope, *Archæologia*, 1908, 257-328.

of both the porch and the hall are enriched by wall arcades. Between the hall and the porch there is an unusual form of wicket-gate or sally port. A small doorway in the hall opens to a short mural passage which leads round the side of the great doors and opens by another small doorway into the porch. At Ludlow the upper stages of the tower were approached by a straight stairway constructed in the thickness of one of the side walls and entered directly from the bailey. It is probable that the outer porch was commanded from above by a machicolation and thus to some extent

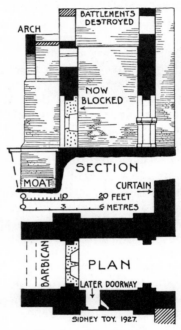

Exeter Castle. The Gateway. p. 103.

took the place of a barbican, but alterations have removed all direct evidence of this arrangement.

The Bâb Al-Futûh, the Bâb An-Nasr, and the Bâb Zuwaylah in the city walls of Cairo were all built in 1085–1091. Each of these gates has a wide gateway, defended by a powerful two-leaved door and by two towers; the towers projecting boldly out on either side of the passage to the door. Between the door and the inner arch there is a large hall with vaulted recesses on either side. The towers of the Bâb An-Nasr are square on the outer face, those of the outer two gates are rounded (pp. 105, 108).

Many examples of twelfth century gateways remain, and among them that at Sherborne Castle, built about 1120 by Roger, Bishop of Salisbury, is one of the most perfect.

FORTIFICATIONS AND BUILDINGS OF THE BAILEY 105

With only a slight projection within the bailey the gateway at Sherborne stands principally outside the curtain wall. It consisted originally of the gateway passage and two upper storeys; the present top storey was added at a later date. This gate was exclusively for the use of the porter and the guard, as the rectangular keep and living quarters within the bailey are of contemporary date. The gateway passage had only one barrier, a two-leaved door placed about a third of the way through from the outside, and there was no portcullis. The flanking walls of the passage are solid except at the end towards the bailey, where there is a small porter's lodge on one side and a spiral stairway to the upper floors on the other. The two upper floors were for the use of the guard, and from the second floor doors open

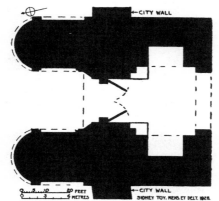

Cairo. Plan of Bâb Al-Futûh. p. 104.

on either side to the walk and battlements of the curtain wall. Recent excavations at Sherborne have brought to light the remains of a succession of gates and connecting passageways projecting out in line from the north wall of the castle.

The gateway of the castle of Newark-on-Trent was built by Alexander, Bishop of Lincoln, 1123–1148. It is of three storeys, was defended simply by a two-leaved door, placed midway in the passage, and there was no portcullis. The upper stages were reached by a stairway built on a spiral vault and entered from the bailey. This gateway appears to have been the keep of the castle.

The south gateway of Launceston Castle was built about 1160, and the gateway to the inner bailey, at Longtown, Herefordshire, about 1180; both are in ruinous condition.

The gateway at Launceston is flanked by round towers which project outside the wall and are built solid for their full height (pp. 78, 106). The gatehouse, which must have extended some 20 ft. within the bailey, has been destroyed with the exception of the outer wall and the flanking towers; it appears to have been two storeys in height above the gateway. The towers projected so far beyond the outer doorway that they formed a narrow entrance, commanded from their battlements on either side. A portcullis, the grooves of which are still to be seen, passed down immediately in front of the doorway. This gateway was defended by a long and narrow barbican, which crossed the moat on low arches and was pierced by loopholes on either side.

At Longtown the gateway is also flanked by solid round towers which project well out from the curtain wall. This gateway is only about half the size of that at Launceston and possibly did not extend into the bailey, consisting only of the existing outer arch with such rooms, now destroyed, as stood above it and the towers. The inner face of the curtain has been

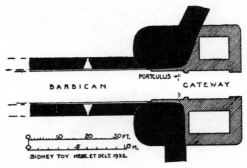

Launceston Castle. Plan of the South Gateway. p. 105.

torn down at this point. At present the only barrier is a portcullis, set about two-thirds of the way through the narrow passage; but whether or not there were any doors beyond the portcullis there is not sufficient evidence to show.

But the most scientifically designed gateways of the latter part of the twelfth century, which remain to us, are those built by Saladin at Cairo 1170–1182. And the finest among them is in that portion of the City walls, on the north-east side of Cairo, known as Burg Ez-Zefer. On the curtailment of the city on this side the portion of the curtain, with its gateway and towers, which was cut off by the new wall was abandoned and allowed to fall into ruin.

The gateway at Burg Ez-Zefer projects entirely on the outside of the wall and has its entrance on the flank, and not on the front face, involving a right-angled turn in passing through it. Close to the gateway, on the entrance side, there is a wall tower, and immediately on the other side of the tower a sally port (pp. 108, 109). The curtain here was defended by

Corfe Castle. Wall walk through the Keep. p. 75.

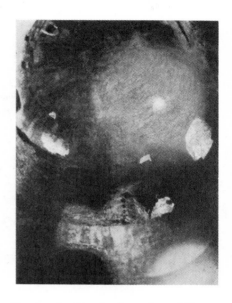

Pembroke Castle. Interior of Keep looking up to the Dome. p. 120.

Colchester Castle. Fireplace in Keep. p. 114.

Rochester Castle. Fireplace in Keep. p. 115.

Cairo. Bâb An-Nasr from without the Walls.   p. 104.

Cairo. Burg Ez-Zefer. The East Gateway from without the Walls.   p. 106.

FORTIFICATIONS AND BUILDINGS OF THE BAILEY

a wide moat and the gateway was approached by a bridge of two spans; the first from the outer bank to a stone pier in the moat, and the second from the pier to a broad stone platform which fills the space between the gateway and the tower. The second part was probably a drawbridge which could be raised or drawn back on the platform. The approach to the entrance was therefore strongly defended; for even if an enemy had gained the platform and had begun to assail the gateway he would find himself, in a confined space, under deadly fire from the meurtrières in the tower at his back and the curtain on his flank, as well as from the battlements

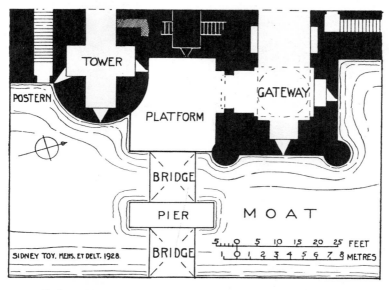

Cairo. Burg Ez-Zefer. Plan of the East Gateway.

on all three sides. He was also open to attack from those issuing from the sally port at his back. The gateway itself was defended by a machicolation, which spans the full width of the opening, and by a two-leaved door secured by a long timber bolt. Piercing the roof above the door there are four holes about 4 in. square and 2 ft. apart. They are on the same vertical plain as the door and their purpose is obscure. But they were possibly connected with some form of iron portcullis, for which there is ample room in the deep tympanum, and which could be dropped in position in the event of the timber doors being broken through or burnt down. This gateway and the curtain in which it occurs is built of strong concrete faced with ashlar.

FIRING LOOPHOLES OR MEURTRIÈRES

The defence of all fortifications in ancient and mediæval times was principally from the battlements of the walls, gateways and towers. But meurtrières, or loopholes, made in the curtain at a level below the battlements, were introduced as early as 215 B.C., as already shown (p. 17). They were described by Philo of Byzantium, about 120 B.C., and were built in the fortifications of Rome in the fourth century and in those of Dara in the sixth century of our era. They do not appear, however, to have been in general

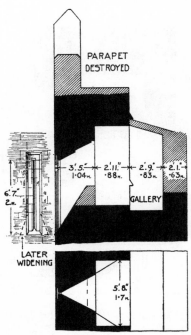

Kenilworth Castle. Plan, section and part elevation of the Battlements of the Keep. p. 111.

use in Western Europe until the twelfth century, and not until the end of that century to have been built in the upper battlements.

Loopholes of this kind enabled the defenders to shoot at the enemy outside the fortress while they themselves remained unseen and safe from attack behind its walls. When built in the walls, each of them consists of the loophole proper, a narrow vertical slot on the outside face of the wall with deeply splayed inner jambs; and a recess in the wall behind the loophole for the accommodation of the archer. The splayed sides of the hole enabled the archer to direct his fire towards the flank as well as the front, and

FORTIFICATIONS AND BUILDINGS OF THE BAILEY

since the sill was deflected steeply downwards from the inside to outside he commanded the ground level also. The recess was often provided with one or two seats.

In the ancient form the outer hole was a simple vertical slot, those made at Syracuse in 215 B.C. being 4 in. wide and 6 ft. long; and this simple form was maintained in the earliest examples of the Middle Ages, though the width of the opening was reduced. Even after other forms were in constant use, and side by side with them, the simple slot was still employed, as at Corfe, about 1270, where the holes are 1½ in. wide and 12 ft. long.

Among the earliest existing mediæval examples are those in the keep at Kenilworth Castle, constructed about 1130. They pierce the walls below the battlements in line with the roof gutter and are approached from the walk which passed round three sides of the keep at this level (p. 110).

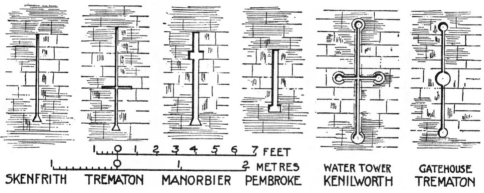

SKENFRITH   TREMATON   MANORBIER   PEMBROKE   WATER TOWER KENILWORTH   GATEHOUSE TREMATON

Loopholes or Meurtrières. p. 112.

There are three on the south side of the keep and two on the west; the east wall supported the gable of the roof and is not pierced, and the north wall has been destroyed. All the loopholes are in their original condition except that in each case, the lower third of the hole on the outer face is cut away so as to form a large triangular-shaped foot 2 ft. 6 in. wide at the base. This cutting was probably done during the civil wars of the thirteenth century in order to give the crossbows, then in use, greater play from side to side. The cutting is roughly executed in the loopholes on the south side of the keep but more skilfully done in those on the west. In each of the last a cross slot was also cut at the same time. In their original condition the slots were from ½ in. to 1 in. wide and from 5 ft. 6 in. to 6 ft. 7 in. long. The recesses for the archers were from 5 ft. to 5 ft. 3 in. wide, by 7 ft. high to the crown of the low arches, and contain stone seats, built across the inner angles of the recesses.

Towards the end of the twelfth century the holes were generally constructed with small triangular feet, and occasionally were bisected by a short horizontal slot giving the whole the form of a cross, as at Skenfrith and Trematon, both of which were constructed about 1190; that at Trematon being inserted in older work (p. 111). The horizontal slots, which were widely splayed at the back, gave the archer a wide lateral sweep for his arrows and bolts, and were introduced especially for the use of the crossbow. The loopholes in the battlements of the keep at Pembroke Castle, built about 1200, were cut square at the base. In all cases the enlargement of

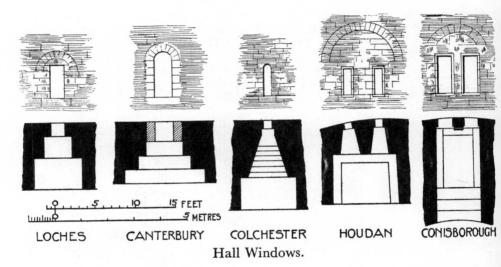

LOCHES   CANTERBURY   COLCHESTER   HOUDAN   CONISBOROUGH

Hall Windows.

the base of the hole was made to give the archer wider range when shooting low.

During the thirteenth century the loopholes were usually terminated both at the base and the head by circular enlargements, and if there was a cross slot that also had a similar termination at either end, as in the water tower at Kenilworth. Sometimes the cross slot was omitted and its place taken by a circular hole, slightly larger than those at the ends, as at the gate-house at Trematon and in the walls of Marten's Tower at Chepstow. But no further alteration occurred in the general design of these holes until the introduction of firearms in the fourteenth century.

WINDOWS

On the first floor of the gatehouse at Exeter are two windows, grouped together, which in design follow the Saxon tradition and are similar to those in the tower of Deerhurst church, built about fifteen years previously.

FORTIFICATIONS AND BUILDINGS OF THE BAILEY

They are carried straight through the wall, have triangular heads and shallow outside recesses.

But generally the windows of the upper or living rooms of the keeps of the eleventh century, judging from existing examples, were from 12 in. to 18 in. wide and about 4 ft. high. They had round or flat external heads, flush with the outer face of the wall, and their internal jambs and round rear arches were either splayed or opened out in orders. The windows were set in wide and lofty internal recesses. Among the best preserved examples are those in the keeps at Loches and Colchester. At Canterbury the upper windows have each a series of three internal recesses, which

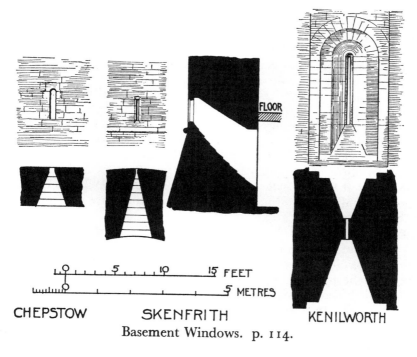

CHEPSTOW   SKENFRITH   KENILWORTH
Basement Windows. p. 114.

diminish in size towards the window opening. Unfortunately the outer openings are so ruinous that it is not possible to say what their original form may have been (p. 112).

In all these cases light was considered of much less importance than safety. The light admitted by the small windows, and through the thick walls of the keeps of Loches and Colchester must have given but poor illumination to the great halls within. But on the other hand, while such windows could be quickly and effectively closed by shutters, even when open the danger from the enemy's missiles was relatively small. It was felt, however, in the twelfth century, that more light to the living apartments was desirable, and the first efforts in this direction, apart from increasing the width of the openings, was to make two openings in one recess; as in the keep at Houdan, about 1130. Here the lights are each 1 ft. wide by 3 ft. 4 in. high and are placed 2 ft. 9 in. apart. The recess behind is very wide and is provided with a seat on all three sides. There were similar windows in the keep at Longtown, about 1180, though here the wide mullions between the lights have been torn away.

At Conisborough, about 1190, the twin lights of the windows are much larger than those at Houdan or Longtown, each of the lights being 1 ft. 10 in. wide by 4 ft. 8 in. high; and since the walls of the keep are very thick the recess behind the lights is correspondingly deep. The lights here are recessed inside for wood shutters and the mullion has a projection in the middle with a hole for the horizontal bar which secured the shutters when closed. Later, when the windows became wider, they were defended by grilles of iron stanchions and saddle bars.

The windows of the ground floors of towers and halls, generally giving air if not illumination to store-rooms, were narrow, single lights. In the early examples the small openings were flush with the outside face of the wall and their inner jambs and rear arches were widely splayed; often their sills were stepped down rapidly from outside to inside. The lower windows of the north tower of the Château de Foix, and those of the old hall in the middle bailey at Corfe Castle, and of the hall built by William Fitz Osbern at Chepstow, all dating from the eleventh century, are perfect examples of their period (p. 113).

In keeps built during the eleventh and twelfth centuries these ground floor lights were mere loopholes, placed so high out of the reach of sappers that they were often above the level of the first floor; their inner lintels and sills being deflected rapidly downwards through the wall to the level of the rooms they ventilated; as at Canterbury, La Roche Guyon, and Skenfrith. During the twelfth century the openings were sometimes splayed on the outside as well as within; as in the fine example in the keep at Kenilworth, about 1130 (pp. 113, 73).

FIREPLACES

Fireplaces of the eleventh century were plain arched openings with semi-circular backs. Their flues after rising up for a short distance within the wall passed through to the outside face and terminated in one or two loopholes; the loopholes being generally concealed in the inner angles of buttresses. The lower courses of the backs of the fireplaces, where combustion occurred, were built of selected stones, often laid in herringbone manner, as at Colchester and Canterbury (pp. 107, 115).

Very little alteration was made in the form of fireplaces from the eleventh century until about 1180. From the early part of the twelfth century the jambs were enriched with small shafts and the arches with chevron mouldings, as at Castle Hedingham and Rochester (p. 107). But although the fireplace sometimes projects slightly out from the wall and has a straight moulding over the head, as at Rochester, there is no real hood. In shell keeps the flues were carried up to the wall walk, a relatively short distance from the upper floors; but in the rectangular keep at Newcastle, as late as 1177, they were still carried through the wall to loopholes on its outer face.

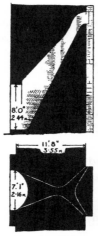

Colchester Castle. Plan and Section of Fireplace in the Keep. p. 114.

Here, however, the semi-circular plan is no longer used, but the fireplaces have straight backs and splayed sides.

The fireplaces in the keep at Conisborough, built about 1190, show a marked development. The flues pass up to the top of the wall, where they terminated in a chimney; and the jambs carry a flat arch or lintel, built of voussoirs with joggled joints. Above the lintel there is a tall stone hood. At Conisborough the hood was a necessity of construction rather than a development in design. For since the fireplace forms a chord across the circular chamber in which it is built, any wall, built on the lintel for the flue, would naturally fall back against the sides of the chamber as it rises. But from this period hoods, projecting well out into the chamber and supported on corbels and small shafts, on either side, were built to fireplaces generally.

CHAPTER XI

CASTLES FROM ABOUT 1190 TO 1270

THE arduous campaigns of the Third Crusade resulted in a further development in military architecture, as well in Western Europe as in Palestine itself. Weak points in some of the existing fortifications had been clearly demonstrated. The Crusaders had seen the great execution wrought by the powerful siege engines on both sides and the dire effects of sapping and mining, and realised that a more scientific plan than that hitherto adopted was essential.

The site now chosen for a new castle, where such choice was possible, was the summit of a precipitous hill; the citadel or inner bailey being backed against the cliff. The main defence was concentrated in the direction of approach, and here there were often two or even three lines of advance fortifications. Château Gaillard, Eure, built 1196–1198; Pembroke Castle, about 1200; and Beeston Castle, Cheshire, about 1225; are all of this order. In the case of castles already built, one or two outer baileys were added on the line of approach, as at Corfe and Chepstow. The living quarters, with the hall and domestic offices, and the chapel were now all built in the court of the inner bailey. The keep, often no longer the ordinary residence of the lord but essentially his last line of defence, is smaller than those built previously but of more powerful and scientific design.

Among the first of these castles was the Château Gaillard, which stands on a precipitous cliff 300 ft. above the river Seine. It was built by Richard I and when complete in 1198 was one of the most powerful castles of the day. The statement that on the completion of the work Richard exclaimed *Ecce quam pulcra filia unius anni* (Behold! what a beautiful daughter of one year!) rests solely on the authority of a chronicle of about 1436, accredited to John Brompton, a work which contains many fables and about which Sir T. D. Hardy said: "There is no reason to believe that it was based on a previous compilation." The building accounts make it clear that the construction occupied three years.

The castle consists of three baileys, arranged in line, the inner bailey being on the edge of the cliff (p. 117). The outer bailey, which was triangular in plan with the apex pointing outwards, was completely surrounded by a moat; and there was a moat between the middle and the inner baileys. The curtains of both the outer and middle baileys were strengthened by circular wall towers. The curtain of the inner bailey has no wall towers, properly so called, but has itself, on the outer face, a continuous series of corrugations, from the battlements of which the faces of the wall could be swept by flank shooting.

The keep, or donjon, stands on the edge of the precipice, principally within but partially projecting outside the inner bailey, where its base rises up from a ledge of rock 40 ft. below the level of the courtyard. It is circular except that the side towards the courtyard is thickened and shaped like the prow of a ship. The prow faced in the direction most vulnerable to attack from the battering-ram and the sapper. It formed a triangular wall the full height of the tower, the outer portion of which might be attacked and partially destroyed without serious effect to the rest of the work. Further, since it pointed directly towards the enemy, its oblique surfaces would deflect without receiving the full force of his projectiles.

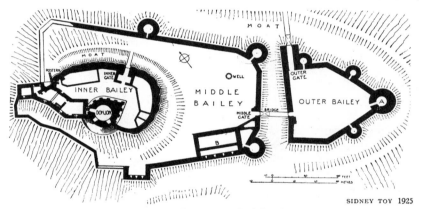

Plan of Château Gaillard.

Additional protection against sapping was obtained by a deep battered plinth and by machicolations, the supports of which rise from the plinth and are carried up the face of the donjon to the first parapet (p. 118). This is one of the earliest examples in Western Europe of stone machicolations at the battlements. At present the donjon consists of two complete storeys; the lower storey being a store-room, and the upper a guard-room. Above this level the walls have been destroyed; but they were apparently only carried up to a height sufficient to screen the roof, most of the corbels of which still remain, and to form the battlements. Having regard to the position of the roof corbels on the inside and the inclination of the buttress-like projections for the machicolations on the outside there can be little doubt but that there were two lines of battlements, one rising above and behind the other.

The entrance to the donjon was by means of flights of steps, now destroyed, up from the courtyard to a doorway in the upper storey. In this room there are two windows, each of two lights, but there are no wall chambers and there is no fireplace. From here descent to the lower floor and ascent to the battlements must have been made by timber stairways, perhaps movable at will, for there is no other provision. In design and construction this keep is of a purely military character, of use only for observation and defence. The hall, living rooms, and domestic offices are built in the courtyard near the donjon; while other structures now represented by foundations only, stood against the curtain in other parts of the inner bailey.

At La Roche Guyon, twenty miles further up the Seine, there is a donjon with a prow similar to that at Château Gaillard, but with plain surfaces.

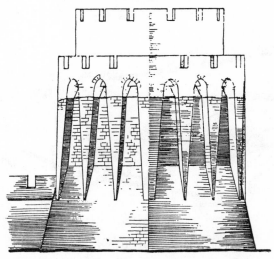

Château Gaillard. The Donjon from the East. p. 117.

Here the donjon stands on high ground, backed on to a precipitous escarpment, while the main buildings of the castle are on the river bank, below the cliff. Ascent from the latter to the donjon was by way of a series of subterranean stairways and narrow ledges cut through and into the rock. All the existing buildings above ground are of a period subsequent to the events recorded by Abbot Suger as having taken place at La Roche Guyon in 1109.[1] The donjon dates probably from the last half of the twelfth century, perhaps a few years earlier than the Château Gaillard, and the château below, in some parts from the same period, but principally from the fifteenth century (p. 119).

The donjon consists of a central tower, a chemise, and an outer wall;

[1] *Vita Ludovici Grossi Regis,* Ch. XVII.

the outer wall enclosing the greater part of the chemise. All three are prow shaped and have their prows pointing in the same direction; away from the cliff and in the line of approach. Though both of the outer walls were of considerable height the central tower rises high above them. In the space between the tower and the chemise there is a well. A large portion of this space was roofed over, forming stores below and a wide fighting platform above, the platform running round the prow of the tower. There were therefore three lines of defence, one from the battlements of the outer wall, the second from those of the chemise, and the third, high above the others, from the battlements, of the central tower.

The central tower consists of a low basement, two upper storeys, and the lower part of what was probably a fourth storey; the rest has been destroyed.

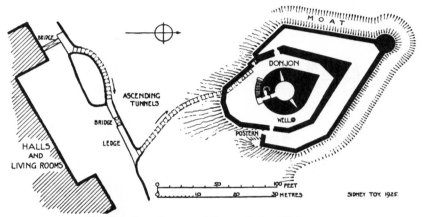

La Roche Guyon. Plan of the Château.

The basement had neither light nor ventilation, and must have been entered through a trap-door in its ceiling. The first floor room received its light through two loopholes, which, externally, are actually side by side with the windows of the room above; but their lintels and sills being deflected steeply down through the thick wall they open out internally to this floor. Here, as also at Château Gaillard, there are neither wall chambers nor fireplaces. But from the entrance passage, which is on the first floor, there is a stairway, built on a spiral vault in the thickness of the wall, which ascended from this level to the upper floors and the battlements.

At Pembroke the inner bailey of the castle, including the keep, was built about 1200 or in the early years of the thirteenth century, and the outer bailey during the first half of the thirteenth century (p. 120). The castle stands on a promontory at the junction of two rivers; the inner bailey being at the head of the promontory. Where the castle is protected by the rivers and their precipitous banks the curtain is relatively thin; but on the side towards the land it is thicker and strengthened at intervals by strong circular towers. One long section of the outer wall was doubled at a later date. The halls and living rooms were in the inner bailey near the keep.

The keep at Pembroke, though by no means among the largest, is one of the most impressive of these round towers. It is 25 ft. 3 in. internal diameter. Its wall is of enormous thickness, and, rising through four storeys

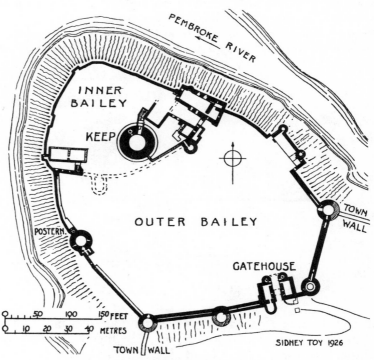

Plan of Pembroke Castle. p. 119.

to the height of about 80 ft., is crowned by a stone dome. There are no windows in the basement and the only light admitted to the first floor comes through two loopholes. Even the second and third floors have only one window each in addition to other narrow loopholes. There is a fireplace on each of the first and second floors[1] (pp. 107, 121, 123).

The entrance was on the first floor and was approached by way of flights of steps, constructed in a narrow fore-building and ending in a drawbridge

---

[1] E. S. Armitage, op. cit. 357, incorrectly says there are no fireplaces. The fireplaces are undoubtedly original.

before the doorway. From the first floor a stairway, constructed on a spiral vault in the thickness of the wall, leads down to the basement and up to the upper floors and the battlements. The wall of the keep is 19 in. thicker on the side of the stairway than it is on the opposite side. At present the keep is entered at basement level by an opening which has been cut through

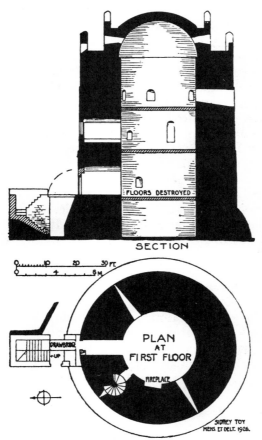

Pembroke Castle. The Keep.

the wall from outside into the stairway; and since a hole, similar to that made for a bolt, occurs in the cutting, it has been suggested that the opening was original.[1] But the opening is clearly not an original doorway. It is roughly driven through an unusually massive plinth, 10 ft. high and spreading 5 ft. at the base. The long horizontal hole probably represents a timber

[1] *Archæologia Cambrensis*, 1883, 203.

tie, such as is often found buried in mediæval masonry and which was cut into when the opening was made.

At the battlements there were two fighting platforms, one in line with the outer face, and the other round the dome behind and above the first. When put in a state of defence, hoarding (p. 205) was built out in front of the lower parapet, and the holes made for the brackets of this temporary work are still to be seen all round the keep. The dome is strongly built at the haunches and is now about 3 ft. 6 in. thick at the crown. The barbican tower, near the gateway of the outer bailey, is constructed with a dome in a similar manner to the keep.

A weak point in the design of many of the earlier round keeps, as in the donjon of Château Gaillard, was that there was only one entrance—or exit. In the event of that doorway being carried by assault the position of those within was desperate; there was no escape. At Pembroke provision is made for such a contingency by a postern on the second floor which led out on to the battlements of buildings near the living quarters, now destroyed.

The round keeps built in France in the early part of the thirteenth century were often vaulted at every storey, or in all storeys but the uppermost, and had two doorways. The Tour du Prisonnier at Gisors, 1206, is three storeys in height and is vaulted in each storey. It is built at the corner of the curtain wall of the old castle and has two outer doorways, one from the wall walk on the west side of the tower and the other from the wall walk on the north side. Both doorways open into the hall on the third storey; and from here spiral stairways lead to the lower floors and the battlements (p. 62).

The Tour Jeanne d'Arc at Rouen, built in 1207, is also of three storeys; but here the interior has been extensively restored and the upper stage rebuilt. The lowest and the middle stages were both, apparently, vaulted. The base of the tower is solid and the floor of the first storey is about 25 ft. up from the ground. At this storey there are two doorways, both reached originally by steps up from the courtyard. The upper stages were gained from this level by a spiral stairway, built in the wall of the tower.

Each of these two towers was at once a residence and a fort, and contained a large fireplace, a well, and a latrine. At Gisors there is also an oven at the back of the fireplace.

At the Tour du Coudray, Chinon, a round donjon of this period, the mural stairways as well as the space inside the entrance doorway are commanded by machicolations, the machicolations passing straight up through the wall to the wall walk (p. 125). The passage from the entrance doorway is not carried straight through into the interior, but, taking a sharp turn to the left immediately inside the doorway, goes along the wall and up some steps to a second doorway on the right. At each turn there

Pembroke Castle. The Keep from the South. p. 120.

Anadoli Kavak. Hieron Castle. The Gateway. p. 88.

Caerphilly Castle. The Screen Wall from the
South-East. p. 155.

is a rectangular machicolation covering the whole space above. By this arrangement an enemy who had forced the doorway found himself checked in the passages and stairways by a deadly rain of missiles from above; while the defenders on the battlements could fight against those within the donjon as well as those without.

The keeps now being built were generally cylindrical. But at Issoudun in France, about 1200, and Araberg, Austria, about 1230, they have prows similar to the central tower at La Roche Guyon; the prow at Araberg Castle being particularly long and sharply pointed. At Ortenberg, Bavaria, dating from the first half of the thirteenth century, the keep is trapezoidal with a prow formed on the shortest side.

Ortenberg Castle consists of three baileys (p. 126). The inner and middle baileys are arranged in line running north and south, the inner

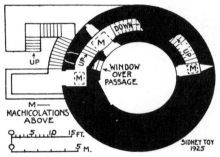

Chinon. Plan of the Tour du Coudray. Entrance Floor. p. 124.

bailey being on the north. The outer bailey occupies a long and relatively narrow space, running north and south in front of the other baileys. Its gateway is at the north end and therefore commanded by, though far below, the inner bailey. There is a considerable rise from the outer to the middle and from the middle to the inner bailey; and the approach is so formed that an enemy entering the gateway would be forced to pass through the whole length of the outer bailey, in the teeth of attack from the inner and middle baileys in succession. Having arrived at the south end of the outer bailey, a turn northward, a long flight of steps, a barbican, and three other gateways had to be negotiated before the inner bailey was reached. The keep stands on the highest point in the middle of the inner bailey, the walls of which so closely surround it as to form a chemise. It commands all parts of the castle inside the walls as well as the approach on the outside.

The formation of approach ways, within as well as without the castle, in such manner that they insured the greatest possible exposure to the enemy and the fullest opportunity of defence to the garrison, was a prominent factor in mediæval defence, and was often carried out with great address, as seen here at Ortenberg.

In Ireland there are several keeps, built during the first half and the middle of the thirteenth century, which are of rectangular plan with a round tower of bold projection at each corner, and three storeys in height; as at Carlow, Co. Carlow; Ferns, Co. Wexford; and Lea, Leix.

Some castles, or inner baileys, of the early part of the thirteenth century

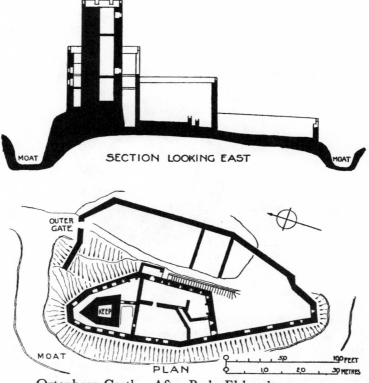

Ortenberg Castle. After Bodo Ebhardt. p. 125.

were built in rectangular form, having a tower at each corner, another in the middle of each of three of the sides, and a gateway near the middle of the fourth side. One of the corner towers, built on a larger scale than the others, was the keep. The castle at Dourdan, Seine et Oise, built about 1220, is designed on this plan, and the inner bailey at Najac, Aveyron, built 1250-60, is of similar character. In the latter, however, the buildings incorporate, at one corner, a square tower built about 1100; and, being oblong in shape, there are intermediate towers on the long sides of the curtain only (pp. 127, 131).

It was now realised that if the defence was to be effective the keep must be not only the place of last resort but the point from which all operations could be directed; and that as well after the enemy had penetrated into the inner bailey as while he was outside the walls. At Najac the curtains all round the inner bailey were provided with an elaborate system of stairways and wall passages, some apparent and others concealed, by means of which the defence forces could be rushed from one part of the fortification to another, as the situation might dictate; and all the operations could be directed from the great round donjon at the south-east corner of the bailey. Each section of the internal defensive system could be isolated from

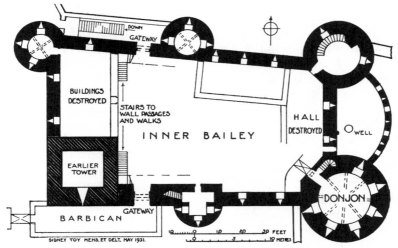

Château de Najac. Plan of the Inner Bailey.

the others by means of barriers; so that if an enemy had penetrated into the courtyard and had carried by assault one section of the curtain, that section could be cut off from the other defences.

Najac stands on the summit of a high hill and is one of the most imposing as it was one of the most powerful castles in the South of France. The donjon is of three storeys, vaulted and pierced with meurtrières at each storey. By the provision of doorways at every stage the means of communication between the donjon and other parts of the castle was complete. There is an entrance at ground floor level, originally reached across a ditch by a drawbridge, and a spiral stairway leads from this level to the upper floors and the battlements. At the second storey, forming the great hall, there are two outer doorways, one leading to a mural passage in the south curtain, and the other to some building in the bailey, now destroyed. At the third storey there are also two outer doorways, one to the battlements of the south curtain and the other to those of the east curtain.

Some round keeps were completely isolated from the curtain by an encircling moat. The keep at Lillebonne, Seine Inférieure, built about 1220, stood isolated at one corner of the castle and was entered by a doorway at ground floor level; the doorway being gained from the courtyard by a drawbridge across the moat. The donjon at Coucy, destroyed by the German armies on their retreat in 1917, was also isolated in this manner, in a position midway in the curtain between two wall towers.

The great donjon at Coucy was at once the largest, strongest and the most magnificent of all mediæval towers. It occupied more than three times the area and was more than twice the height of the keep at Pembroke. It was built during the second quarter of the thirteenth century.[1] Internally it was divided into three storeys, all magnificently vaulted, and was a complete residence in itself; having a well, fireplaces and latrines. Light was admitted into the interior through a large circular eye in the vault at each stage. The entrance, defended by a drawbridge, a machicolation, a portcullis and two doors, was on the ground floor, and there was a postern on the first floor. The postern opened out to another drawbridge, thrown at great height across the moat to a chemise, the chemise passing round, in semi-circular form, on the outside of the moat and connecting the curtain on one side of the donjon to that on the other side. Here the whole of the active defence was conducted from the battlements, where there were firing loopholes and provision for hoardings all round the donjon.

The corner towers at Coucy were designed with very great skill. Circular throughout externally, each of the stages above the basement was hexagonal internally and had wide recesses, greatly adding to the size of the room, on all sides. The plan of each of the upper storeys was so disposed that its sides, and therefore its recesses, occur above the angles of the stage below, so that the meurtrières, set alternately, had the greatest possible range, and the even distribution of its loads added enormously to the resistance of the tower against sapping. The basements are circular, have no recesses, and their thick walls are strengthened by deep plinths.

At Aigues Mortes there is an enormous round donjon, the Tour de Constance, which stands isolated by an encircling moat at one corner of the fortifications and was built about the middle of the thirteenth century. Here the donjon was originally a castle in itself; for the town walls, whether or not they entered into the original design, were not built until the last quarter of the thirteenth century.

The Tour de Constance consists of two stages of large vaulted halls and a small basement; it is surmounted by a tall turret and beacon which were added about 1300 (p. 129). The walls of the tower are 19 ft. thick and are very plain, being pierced only by meurtrières and two small windows. Light is admitted to the interior halls and the basement through large

---

[1] *Le Chateau de Coucy*, by E. Lefevre-Pontalis, p. 49.

CASTLES FROM ABOUT 1190 TO 1270 129

circular eyes, one in the centre of each vault, the uppermost piercing the flat roof of the tower. The entrance and the postern, the first on the south and the other on the north side of the tower, are both at first floor level,

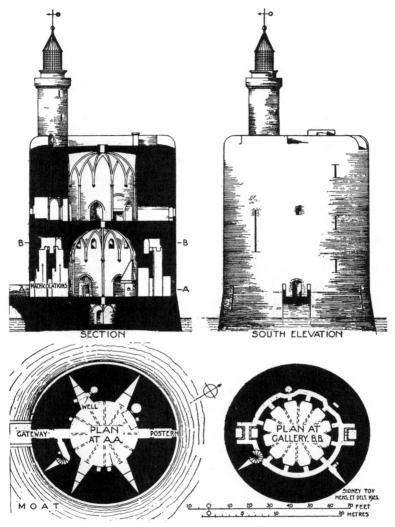

Aigues Mortes. Tour de Constance.

and each of them is defended by a portcullis, two lines of doors, and the postern by one and the entrance by two machicolations. Even if an enemy had passed these barriers and had entered the lower hall he would still be under heavy fire on all sides from the openings of a mural gallery, which is carried round the hall 22 ft. above the floor level, and from the hall above through the eye in the vault. The battlements of this tower were adapted later for the use of artillery.

At Bothwell, Lanarkshire, the defences centred round a great circular keep, placed at one corner of a pentagonal bailey. Bothwell Castle stands on a promontory formed by a sharp loop-shaped turn in the Clyde, and was defended by precipitous banks on the sides towards the river and by a ditch on the land side. It was built in the third quarter of the thirteenth century, and the inner half of the keep is of that period; the outer half, from base to parapet, was thrown down in 1314 when the whole castle was laid in ruins. In 1336 Bothwell was again put in a state of defence and extensive works of rebuilding undertaken. A new wall was built from east to west across the middle of the bailey and the portion of the castle south of the line only was retained and refortified. The gap in the keep was closed by a straight wall, built across the middle of the tower. In a siege of 1337, though great damage was done to some other parts of the castle, the keep appears to have suffered little further injury.

The keep, called Valence Tower, stands on the extreme point of the promontory, commanding the river in both directions. It was 65 ft. in diameter externally, was octagonal internally, and the remaining portion still rises to the full height of four storeys; the first storey being a basement below the level of the bailey (p. 137). A deep moat on the bailey side isolated the keep from the other buildings of the castle. The entrance is at the second storey and is skilfully placed near the curtain at the north end of the moat. Here a pointed projection stands out from the keep and the doorway is set on that side of the projection next the curtain, so that the approach is in line with and close against the curtain, and is commanded by the wall walk above. The doorway was defended by a drawbridge, across the moat, a portcullis, and a door; the drawbridge and portcullis being operated from a small vaulted chamber over the entrance passage. Above the chamber the projection from the keep terminates in a pyramidal spur.

The entrance passage, taking a turn to the right, leads into a fine hall, having a large window to the courtyard, enriched with jamb shafts. On the walls, rising from corbels at the angles, and forming an arch on each side of the octagon, are labels which at first sight suggest vaulting, but on examination appear to be decorative arches and not wall ribs. The floors were of timber; the first, and probably the second, being supported in the centre by an octagonal pier, the lower portion of which still remains in the basement. The first floor was also supported on stone arches thrown across the keep diagonally from the central pier to the walls on either side. A doorway in the south side of the hall opens to a passage leading to a

Château de Najac from the East.   p. 127.

Beeston Castle, from the West.   p. 136.

Clifford's Tower, York, from the North-East.   p. 133.

latrine and another, on the north side, to a spiral stairway leading down to the basement, where there is a well, and up to the upper floors and the battlements. The third storey, relatively plain in character, was probably the military quarters, and the fourth storey, with a large window of two trefoiled lights to the courtyard, the chief living room. Both the timber floor between these storeys and the roof were supported on wall posts and wide spreading struts, and the long chases for the former are still be to seen in the angles and sides of the remaining walls.

The parapet has been destroyed, but on the north side of the keep at this level, high above the entrance doorway and the wall walk of the curtain, are four boldly projecting corbels, spaced from 2 ft. 2 in. to 2 ft. 11 in. apart. These corbels were for the support of hoarding, two of the spaces between them commanding the entrance to the keep and the third the exit from the spiral stairway to the wall walk of the curtain. There are no other corbels at this level in the remaining portion of the keep.

The wall walks on the curtains leading north and south from the keep were protected by parapets both on the side towards the bailey and on that towards the field, and were covered by high pitched roofs, the lines of the verges of which remain on the wall of the keep. The doorway to the wall walk on the south side led out from the keep itself, that on the north side from the spiral stairway. The latter was closed against the keep and secured by a strong timber bar; so that in the event of the entrance to the keep being forced its defenders could escape on to the wall and bar the door against the assailants. There are two posterns in the walls near the keep, one on the north, from the moat, and the other on the south.

This great tower at Bothwell was one of the finest circular keeps in Britain, it had its own water supply and residential requirements, was powerfully fortified, and could be held as well against a foe within as without the castle.

Clifford's Tower, York, standing on the mound of an earlier castle, is a quatrefoil-shaped keep resembling the donjon at Etampes (pp. 132, 134). There can be little doubt that it belongs to the thirteenth century, though the writer of the "official guide" to the tower ascribes it, largely on account of its shouldered arches, to the fourteenth century. Shouldered arches occur in the twelfth century, as at Provins, and in the early part of the thirteenth century, as at Coucy. The chapel, obviously of mid-thirteenth century date, has suffered from violence and distortion, and two of its walls were rebuilt in the seventeenth century. The arcade against the keep was probably set in position after the wall behind it was built, and both it and the arch at the west have been altered. But the east arcade, now partly buried at the south end, is clearly contemporary with the wall in which it occurs and there is no reason to suppose that the keep is of later date than the chapel.

This keep was built probably in 1245-1259, when a sum of nearly £2,000 was spent on the castle. The details, not only of the chapel, but of the construction of the base of the turrets, large corbels instead of a series of corbel courses forming a conical bracket, as at the outer gate at Harlech and the water gate at Beaumaris; and of the loopholes, originally long slots one inch wide and having a circular enlargement at the foot; are also unquestionably of this period. The square openings at the head of the loopholes are not original but were formed at a later date to give light to the interior, which must have been very dark hitherto; the only original windows being two with pointed heads on the upper floor and now partly blocked.

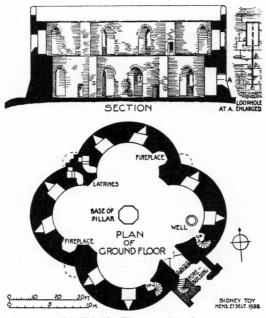

Clifford's Tower, York.

Clifford's Tower is two storeys in height, and is entered at ground floor level through a forebuilding. The floor of the upper storey and the roof, both now destroyed, were supported in the centre by a pier, the foundations of which were found about thirty years ago. Both storeys are pierced all round by meurtrières, or loopholes, placed like those in the wall towers at Coucy, so that the loopholes of the upper storey are in vertical plane midway between those of the ground floor. Two spiral stairways lead from the ground to the upper floor and the battlements, and two other stairways, formed at the re-entering angles of the lobes of the tower, rise from the upper floor to the battlements. The latter are built in small turrets which project on the outside and rise from large corbels at upper floor level. The forebuilding is of three storeys, the first being the entrance porch, the second the chapel with richly moulded arcades, and the third a small chamber from which the portcullis was operated.

At the castle of Angers the strength of the fortifications was concentrated largely on the curtain walls, built 1228–1238. The walls are constructed of slate bonded at intervals by courses of dressed sandstone and granite, the alternate arrangement giving them a strong banded effect. They are of great height and strength and are defended by powerful towers, placed closely together. Both walls and towers have widely battered plinths rising to about half their height and the rock at their base is scarped to their contours. Not only are these walls powerful in themselves but, despite

Angers. The Château from the South.

the fact that most of the towers have lost their upper stages, the general effect to-day is most imposing and awe-inspiring.

In many castles designed in the early part of the thirteenth century the keep was omitted entirely, reliance being placed on the strength of the fortifications as a whole. The powerful castle of Le Krak des Chevaliers, built by the Knights Hospitallers in Syria, is of this period and character.

Le Krak des Chevaliers stands in the mountainous district north of Tripoli. It consists of two baileys, one within the other, which gradually expand in width from the north, where the site is protected naturally by the precipitous fall of the hill, to the south, where the position is more vulnerable. There are three powerful towers in the south wall of both baileys, one at either end and one in the middle; and, since the inner bailey stands high above the outer, its battlements dominate the whole castle both without and within. The entrance to the castle was strongly defended at every point.

The outer gate is placed near the north end of the east wall, so that an approaching enemy would be required to cover about two-thirds of the length of the castle, under direct attack from its walls, in order to reach the first gate. The outer gate being passed, the way through to the inner bailey is by means of sinuous vaulted passages, barred at intervals by gates and commanded from above by machicolations. Finally the inner gate was

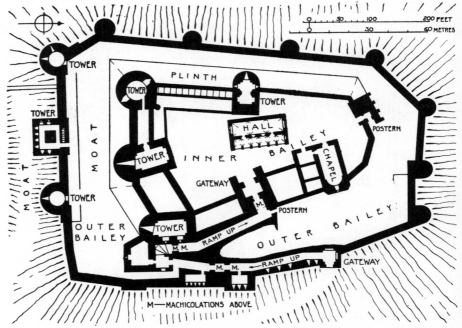

Plan of Le Krak des Chevaliers. After MM. Rey and Guillaumet. p. 135.

defended by a large square machicolation, the full width of the passage, a portcullis, and a door.

Beeston Castle, Cheshire, built about 1225 and now a shattered ruin, appears to have had no keep. It is perched on a high hill with sheer precipices on three sides and a steep slope on the fourth (p. 132). The strength of Beeston lies largely in its inaccessibility. The castle consists of an inner bailey, on the summit of the hill, and a large outer bailey on the sloping ground which stretches away to the east. The curtain of the inner bailey was defended by three wall towers; two of them, with the gatehouse, being concentrated on the east side, away from the precipice and towards the

Bothwell Castle. The Keep from within the Bailey.   p. 130.

Beaumaris Castle. The North Gatehouse from within the Bailey.   p. 161.

Harlech Castle from the North.   p. 160.

Criccieth Castle from the North-West.   p. 161.

outer bailey. Further protection on this side of the inner bailey is secured by a ditch about 30 ft. deep and 35 ft. wide, which has been cut through the natural rock from one precipitous side of the hill to the other. The excavation of this artificial ravine two hundred years before blasting operations were known was a work of great engineering skill.

The Château de Boulogne, completed in 1231, has no keep, but is built on a polygonal plan, with towers at the corners and a gateway, flanked by towers, on one of the sides. The internal buildings are ranged against the curtain round the courtyard. This castle was surrounded by a moat and was defended from meurtrières opening out of tiers of mural passages in the curtain, as well as from the battlements.

Of the same order also are the castles of Maniace, at Syracuse, and Ursino at Catania, built by the Emperor Frederick II in Sicily during the first half of the thirteenth century. Both these castles are designed on a square plan with towers at the corners, and, in the case of Ursino an intermediate tower in the middle of each side. The internal apartments are built against the curtain round the courtyard. At Ursino, communication between the lower and upper floors is secured by spiral stairways, constructed in two of the towers, the other six towers being devoted to the accommodation of chambers. At Maniace, where there are no intermediate towers, there are four stairways, one in each of the corner towers. This castle was protected all round by an outer wall.

The principle of defence by double curtains, as adopted at Le Krak des Chevaliers and Maniace, was applied with great address, in the thirteenth century to the fortifications of Carcassonne. Before that period the fortifications of Carcassonne consisted of a single line of defences and a citadel or castle; the curtain and wall towers dating from Roman and Visigothic times, repaired about 1130, and the castle from the twelfth century. The weak points of these defences having been demonstrated during the Albigensian War, and in the siege of 1240, extensive additions and repairs were begun in 1247 and continued for about forty years.

The existing curtain was extended at one end, and many of the towers, and the gateways, were rebuilt; six of the new towers, including those flanking the east, or Narbonnais, gate, having pointed beaks on the outer face. In addition, a second curtain, with wall towers, was built all round the fortress, outside the first curtain; leaving a relatively narrow terrace called the "lists" between the two lines of fortification. At the north end of the town this terrace is commanded by a large and strong wall tower, which projects partly across it, and at the south end, at a point where the curtains approach closely together, by a tower which is built on both walls and stands astride the lists; the way through the south tower being guarded by barriers and machicolations, so that in the event of the lists being carried by assault, its circulation by the enemy was checked at these two points. On the east side, where the fall of ground is more gradual than on the west, a wide moat was dug before the outer wall, from the north end of the town to the south.

The castle also was strengthened and a strong barbican built before it, the barbican extending down the hill beyond the walls and ending in a large circular outwork. In addition to the two gateways, one on the east and the other on the west side of the town, which pierced both curtains, there were six sally ports in the inner curtain and at least one in the outer curtain. The sally ports are placed in obscure positions and so high in the walls that they could be reached only by means of ladders.

In these works every known artifice and perfection of military architecture was employed, every contingency provided for. When complete, about 1285, Carcassonne was a formidable fortress well deserving its widespread renown.

CHAPTER XII

SIEGE ENGINES AND SIEGE OPERATIONS OF THE MIDDLE AGES

THE siege engines used and the methods of attack and defence employed in ancient and classical times have been described (pp. 13, 18, 27, 40). During the Middle Ages and until the effective use of gunpowder in the fifteenth century, similar engines, developed on more powerful lines, were in use; and the same methods of sapping and mining were employed. There is abundant evidence to show that as the barbarians came into contact with civilized nations the former studied and adopted the more scientific methods of warfare. Cæsar found the Germanic tribes particularly apt in this respect, though their first attempts were crude.[1] At the siege of Rome, A.D. 537, the artillery brought up by the Goths inspired wonder and terror in the hearts of the citizens themselves, as, standing behind the battlements, the Romans watched the advance towards the walls of powerful siege engines and towers from all sides.[2] Again, when in 885 Rollo laid siege to Paris, it was with the siege engines of his day.[3]

Among the hand weapons in use during the Middle Ages the bow and arrow still held a strong position, and that long after the introduction of the crossbow. The crossbow came into prominence in European warfare early in the twelfth century. It consisted of a wood stock, similar in form to the butt of a musket, and a bow fixed to one end of the stock. In the earlier forms the bow was made of wood, or a composite of wood, horn, sinew and glue, but after about 1370 it was made of steel. The bowstring was stretched by means of a lever or small windlass, and the bolt, or quarrel, was released by a trigger. The wounds inflicted by this weapon were considered be to so barbarous that its use was proscribed by the Lateran Council of 1139. But notwithstanding this prohibition the crossbow was in general use at the end of the twelfth century and, except among the English, was the favourite weapon from that time to the latter part of the fifteenth century. In open warfare the English preferred the longbow, which was about 6 ft. long. The longbow was light while the crossbow was heavy and cumbersome. With the longbow the archer could shoot about five arrows while the crossbowman was discharging one bolt, and he could keep his eye on the foe during the adjustment of a new missile while the crossbowman's whole attention was required for this purpose.

---

[1] Tacitus: *The Histories*, Book IV, 23.
[2] Procopius: *History of the Wars*, Book V, C. XXII.
[3] *Guillaume de Jumiège:* Book II, C. XII.

In the defence of fortifications, however, where the crossbowman would have support for his bow and himself be secure from attack, the crossbow, with its heavier missile, greater force, and longer range was by far the superior weapon. The effective range of the longbow was about 220 yards, that of a fifteenth-century crossbow was from 370 to 380 yards, and with some bows even greater. In 1901 Sir Ralph Payne-Gallwey, using a crossbow of the fifteenth century with a steel bow, shot several bolts across the Menai Straits at a point where the distance was from 440 to 450 yards.[1] Much longer ranges have been claimed.

Scaling ladders were used at all periods, some of them being made of thongs so that they could be thrown over the walls.[2] Battering-rams were built within strong timber houses, which were covered, as a protection from fire, either with iron plates[3] or raw hides, and were mounted on wheels. They were brought up to the walls by teams of men working from the inside and propelling them along by means of poles. When in position the wheels were removed and the machine fixed by means of wooden pegs.[4] The ram had an iron head and was swung to and fro by picked men working on either side. Rams were cumbersome machines, and, working under constant exposure, were often destroyed, while their heads could be grappled and held by chains let down from the walls. They were used in the early crusades, but were being gradually superseded by powerful trebuchets and other projectile engines. They did not appear to have been in the equipment of King Stephen, whose siege engines played an important rôle in his campaigns throughout England in the middle of the twelfth century.

Siege towers, or beffrois, many storeys in height, were also built of timber, covered with raw hides and mounted on wheels. Their great height enabled the besiegers to fight on a level with those on the walls, or even the towers, of the castle assaulted. When brought up close enough, a bridge was thrown across from the tower to the battlements of the castle and those in the tower rushed across it on to the walls, while others passed up through the tower and on to the bridge in a continuous stream.

In addition to penthouses as used by the Romans, there were also cats, or mobile penthouses. The cat was a long one-storey structure, built of stout timbers and covered with raw hides; it was brought up in position either by means of rollers and levers or by a system of pulleys and windlasses. Arrived before the walls, men working under its protection built a causeway across the moat, and when finally it was moved over the causeway, it formed a secure shelter for those sapping the base of the wall. Mantlets, or wood

---

[1] *The Crossbow*, by Sir Ralph Payne-Gallwey, 1903, p. 14.
[2] *Acts of Stephen*, Book I.
[3] *Geoffrey de Vinsauf*, Book I, C. LX.
[4] *Anna Comnena, Alexias*, Book XIII.

screens, protecting small bodies of archers, were placed in convenient positions before the fortress assailed.

Projectile engines, worked by means of springs, thongs, twisted ropes, or counterpoised weights, have been given various names, sometimes interchangeable. They may be grouped under three heads—petrariæ, engines casting huge rocks of stone; ballistæ, or mangonels, for stones of about a half cwt.; and catapults, or scorpians, for casting smaller stones, darts and firebrands. The engines used at Syracuse in 215 B.C., and at Jerusalem A.D. 70, have been described above. By the end of the twelfth century projectile engines had become almost as powerful as early cannon. At the siege of Acre in 1189-91 the King of France had a petraria, called "Bad Neighbour", which, by constant blows, broke down part of the main wall of the city. At the same siege, one of the engines belonging to King Richard of England killed twelve men at one shot. This latter incident astonished the Saracens so much that they brought the stone ball to Saladin for inspection.[1]

As far as is known none of these powerful engines has survived. Drawings from contemporary data have been made by M. Viollet-le-Duc and others. But with the exception of a sketch of one part of a trebuchet in the *Album* by Villard de Honnecourt, a thirteenth-century architect, there are no reliable contemporary illustrations. A trebuchet was a powerful projectile engine, worked by springs and counterpoised weights, and the sketch illustrated its framed sole plate. If there were sketches of other parts of the engine in the collection they are among those which have been lost.

The missiles used included stones, darts, poles sharpened at the points, and firebrands. Fire was always one of the chief weapons used; flaming torches, burning pitch, and boiling oil were thrown from the walls on the besiegers; and burning and highly inflammable missiles were projected from the engines of both parties. Greek fire (p. 49), having the property of spreading in all directions, was thrown from the engines of the Saracens on the Crusaders to their great terror and consternation.

When direct assault on a castle had failed attempt was made to bring down the walls either by sapping their bases under the protection of a penthouse, or by mining below their foundations. Mining was often effective in reducing a fortress, and the only defence against it was countermining. During the Middle Ages both operations were conducted with great skill and address.

Siege operations are best understood and the whole purpose of mediæval defences appreciated by a perusal of the very vivid descriptions of sieges given by contemporary chroniclers.

---

[1] *Geoffrey de Vinsauf,* Book III, C. 7.

SIEGE OF NICÆA, 1097

On their march towards Jerusalem the armies of the First Crusade arrived before Nicæa in 1097 and laid siege to the city; surrounding those parts of the walls not washed by the shores of the lake. After they had scoured the neighbouring forests for timber for the construction of their siege engines and had set up their artillery they began their attack on the walls; hoping to accomplish their object both by famine and assault. Finding that the citizens were receiving supplies by way of the lake, the crusaders, with the help of Alexios, Emperor of Constantinople, transported a fleet of ships, on waggons, seven miles overland in one night; launched them on the lake, and so cut off that means of succour.

Meanwhile the engines, pounding away at the walls, met with vigorous resistance from the garrison, who threw torches as well as pitch, oil, lard and other inflammable materials at them, destroying a great number. One large tower offered great resistance, and after incessant attack by two siege engines not a stone was moved. At length, when more engines were brought up against it and larger and harder stone missiles employed, some fissures were made in its wall. The base of the wall was attacked by a battering-ram and by sappers with crow-bars. But all these efforts were in vain, for the breaches made during the day were repaired by the garrison during the night. Eventually a very strong pent-house was built and brought up to the tower. Under the protection of this covering, on which flaming materials and huge rocks of stone were thrown without effect, sappers worked away at the base of the wall.

As the masonry at the base was removed by the sappers its place was supplied by props and stays, and when a cavity of sufficient size for their purpose had been made, combustible materials were thrown among the timber work. The men then set fire to the props and escaped back to the camp, leaving the pent-house where it was. At midnight, when the props were consumed, the tower fell with such a deafening crash that the sound could be heard from a great distance. After that event the citizens, realising that their case was hopeless, surrendered themselves to the Emperor Alexios and the city was taken.[1]

SIEGE OF ANTIOCH, 1097–98

Antioch at this period was a city of great extent, about three miles long by two miles wide, and was bounded on the north by the river Orontes. It stands partly on the plain and partly on the slopes of a high hill, the citadel being situated on the highest point on the southern boundary. On their arrival at Antioch from Nicæa the crusaders proceeded to attack the city. Investment of such extensive lines and difficult ground was probably deemed

---

[1] *Anna Comnena,* Book XI, and *William of Tyre.*

impracticable, and the armies disposed themselves before the principal gates and built four forts.

Here the crusaders met with strong and effective resistance. The garrison was under the command of a skilful leader who had powerful engines mounted on the walls and was thus able to repel the enemy's attacks. Destructive sallies were also made from one of the gates, and to counter this manœuvre the crusaders blocked the gate by rolling great rocks and heavy logs of oak against it. It was only by the treachery of one Emir-Feir that the city was eventually taken. This man, having killed his own brother, introduced some of the crusaders on the battlements by means of a rope ladder; they, having gained a postern, let in the others, and then, all proceeding to the main gate, let in the rest of the army.

Once within the gates a dreadful massacre ensued, in which neither sex nor age was spared. Demanding the names of the most important houses they entered them, slew the domestics, and, penetrating into the private apartments, transfixed alike nobles, mothers of families, and infants. Pillaging everywhere, they carried away rich vestments and vessels of gold and silver.

An incident during this siege might be recorded. The report had spread that there were spies among the army of the crusaders, and at a council of leaders Bœmond undertook to deal with the matter. One night when the usual preparations for supper were being made Bœmond brought some Turks out of prison, sent them to the slaughter-house, ordering that their throats were to be cut and their bodies roasted, prepared, and carefully laid out as for eating. He also gave instructions that if any enquiries were made as to the meaning of these proceedings, the answer to be given was that the princes had issued orders that henceforth all prisoners and spies were to be treated in the same manner and eaten by them and their people. The report which thereupon spread abroad that the invaders were a people of abnormal cruelty, who not only imprisoned and tortured their enemies but also ate their flesh and drank their blood, sent a thrill of horror throughout the land.[1]

SIEGE OF JERUSALEM, 1099

In the following summer the crusaders appeared before Jerusalem and, having disposed their forces round the city, set up their engines in convenient positions and began the assault. The Turks also set up engines on the walls, noted the construction of their enemy's machines that they might copy the designs, and returned the fire with great vigour. Flaming torches, rope dipped in sulphur, oil, pitch, fragile pots filled with inflammable materials and breaking easily on impact, and all sorts of similar missiles were thrown from the Turks' machines on the engines and siege towers of the crusaders, in order that they might ignite and destroy them.

The crusaders built three particularly strong siege towers and moved them

[1] *William of Tyre.*

up to the walls at three different points; one of them being brought up piece by piece during the night and reassembled in its new position. Each of the towers had a drawbridge on one side which, when close enough to the wall, could be let down to the battlements and so form a way for the troops into the city. From the summits of these towers men with ballistæ, bows, and other weapons poured a rain of blows on the battlements opposite them. The towers were met with shows of missiles and so effective was the Turkish fire that those occupying the upper parts were rendered dizzy by the constant shaking, while the towers themselves began to spread at the base.

Among the engines of the crusaders there was one which threw enormous stones with unusual force. It did great execution among those on the battlements and the enemy's attacks on it had no effect. The Turks then brought up two witches and set them on the wall in order that they might curse it; but a missile from the engine struck and killed both the witches as well as three other women who were with them. However the defence of the city was so well conducted and the fire on the engines and towers of the besiegers so effective that the crusaders became despondent and considered drawing off the attack, but eventually decided to renew their efforts.

In order to deaden the impact and shock of missiles, the Turks had suspended outside the walls sacks of straw and tow, cushions, carpets and beams of wood. The crusaders on the north side of the city, having filled the ditch and brought up one of their towers close to the wall, cut down two of the suspended beams. They then set fire to the sacks of straw, mattresses, and other combustible materials suspended before the walls, and the wind, blowing from the north, drove such dense volumes of smoke into the city that those on the walls could neither open their mouths nor their eyes and had perforce to beat a retreat. The crusaders then let down their drawbridge on the wall, using the timbers they had cut down as sleepers, thrown across between the tower and the battlements, on which the drawbridge rested and by which it was greatly strengthened. The forces on this side then entered the city and, since the wall was forsaken here, used ladders as well as their tower and bridge. Having entered, they opened one of the gates and the troops poured *pêle-mêle* into the streets within. Meanwhile, the forces on the south side, hearing that the city had been taken, applied ladders to the wall and entered without opposition.

Now again a scene of the most awful massacre and butchery ensued. The troops rushed through the city killing everywhere, sparing none, man, woman, or infant, dragging out those who had hidden themselves into public places and slaying them like beasts. In this scene of carnage and blood the forces from the north and those from the south met in the centre of the city.[1]

---

[1] *William of Tyre.*

SIEGE OF ACRE, 1189–1192

Acre was defended on one side by the sea and on the other by three lines of fortifications and a citadel within the third line. On the south there was a harbour, protected by a mole and a strong tower, called the Tower of Flies, at the end of the mole. On their arrival at Acre in 1189 the armies of the Third Crusade began their attack on the city both by sea and land.

The fleet, having erected on their galleys a tall siege tower and other engines, all covered with raw hides, made a vigorous attack on the Tower of Flies. Those in the tower, assisted by the citizens who came to their aid, responded with equal energy; they threw Greek fire on the siege tower, and on the other machines of their foes, and by this means destroyed them, and so the attack from this side failed.

On the land side the crusaders were more successful. They first fortified their own camp and then threw up an entrenchment in front of the walls of the city from shore to shore. All supplies now being cut off the Turks, stricken with famine, offered to surrender on condition that they should be allowed to depart with their property unmolested. This condition was refused and the siege was protracted for two years.

The crusaders constructed three siege towers, which were built up in storeys and reached to a greater height than the walls of the city. Twisted ropes were hung in front of the towers to deaden the force of missiles. There were also battering-rams, one of which was covered with iron plates, and a large number of powerful projectile engines; the latter not only attacking the walls but covering the advance of the towers. But the Turks possessed no less powerful engines and made valiant resistance. Their machines cast stones of immense weight at great distances, destroying everything they struck. They broke in pieces some of the besiegers' petrariæ and rendered other machines useless. Speaking of the defenders the chronicler writes: "Never were there braver warriors of any creed on earth; and the memory of their actions excites at once our respect and astonishment."

Recourse was then had to mining. At one point the French made an attack on a strong corner tower called "The Cursed," and by diligent digging made a cavity, supporting its roof with logs of wood. But the Turks by counter-mining reached the same spot and frustrated their designs. At another point Richard, King of England, made an attack on a tower both by mining and projectile engines; with the result that the tower, or its outer wall, was brought down. But when the men tried to rush through the breach they were repelled. Then a company of Pisans ran forward and they were driven out also; the Turks fighting with their swords and with Greek fire.

Eventually, their fortifications partly destroyed and themselves greatly reduced in numbers, the Turks submitted to severe conditions of surrender, many of which were beyond their power to perform. Even so, as they departed penniless from the city their spirit and courageous bearing struck admiration in the hearts of all who saw them.[1]

SIEGE OF CHATEAU GAILLARD, 1203–1204

The strength of Château Gaillard was put to a severe test five years after it was built, when it was attacked by the King of France and held for King John of England by Roger de Lacy. In 1203 Philip advanced towards the castle and after desperate struggles on both sides took the town of Les Petit Andeleys, which King Richard had built on the Seine at its foot. Having regard to the strength of the castle Philip decided to starve the garrison into submission; and with this end in view he dug two lines of trenches, running from the water at the base of the hill on which the castle stands to the top of the hill and from there down to the river on the other side. At intervals in the space between the trenches he built timber towers and placed guards not only in the towers but also all along the intervals between them. The area enclosed between these lines of fortification and the river included not only the castle but also the little valleys surrounding it. Philip's troops then sat down for about three months to await events; and thereupon ensued one of the most terrible episodes in the history of the Middle Ages.

On the hillside between the town and the castle there was a thickly populated street of houses whose inhabitants, when the town was taken, retired up to the castle and were received within its walls. But as the siege was prolonged one thousand of them were sent out and were allowed to pass the French lines. Later, as the reserve supplies were being consumed, de Lacy, selecting those most useful to him, sent forth all the others, to the number of four hundred, including infirm men, women, and children. They went out with joy; but when the gate was closed behind them they were met with a volley of missiles from the French, who had been ordered to allow no others to pass. Then, repulsed on both sides and under continual attack, they found themselves confined to the valleys between the castle and the French trenches. There they remained during a severe winter suffering intensely from want, hunger, and exposure, endeavouring to sustain their existence on such winter herbs as they could find. Dogs which had been driven out of the castle were seized and eaten, the skins as well as the flesh. A new-born infant was immediately devoured. Their condition was so deplorable that at length, after three months of intense suffering those of them who survived were allowed to pass through the French lines; but nearly all of this remnant died on taking food.

In the spring of 1204, following these events, Philip set up his siege engines on the high ground to the south-east and began his assault on the castle (p. 117). His engines included petrariæ, mangonels, mantlets and a very high siege tower. He also built a long pent-house for the protection of those en-

[1] *Geoffrey de Vinsauf.*

SIEGE ENGINES AND SIEGE OPERATIONS 149

gaged in filling the castle ditches. To his attack the garrison replied vigorously with stones from their own engines, causing considerable loss to the besiegers. The French then began sapping operations, under the protection of their shields, on the salient tower of the outer bailey A and were successful in excavating a cavity, strutting, firing the timber, and in bringing down the tower. The outer bailey was then taken and the French proceeded to attack the remaining works.

Against the curtain on the south side of the middle bailey there was a building B, which had latrines in its lower storey and contained a chapel in the upper storey; the chapel having a window in the outer wall. A French soldier, observing this window, he, with some companions, made endeavours to reach it. They searched along the river bank for the outlet of the drain from the latrines, found it, crawled up through the drain, and gained a point just below the window of the chapel. Here, mounting on the shoulders of one of his companions, the soldier sprang up to the window and by means of a cord brought the others up also. Having got within they began to make a great noise and the garrison, believing that a great number of the enemy had entered, set fire to the whole building and retired within the inner bailey. Those who had entered, however, were able to protect themselves in the vaults, and before the fire had died down they rushed out, lowered the drawbridge between the outer and middle baileys so that the French troops could enter; and the middle bailey was taken.

Then the French, under the protection of one of their machines, proceeded to undermine the wall of the inner bailey. But by countermining on the other side the garrison broke into the French tunnel and drove them out. The wall, however, weakened by being undermined on both sides, and attacked by a powerful petraria, throwing enormous blocks of stone, was fractured; and the French, rushing through the breach, entered the inner bailey. Even then, none of the garrison surrendered but all fought as long as it was possible to do so.[1]

SIEGE OF ROCHESTER CASTLE, 1215

Rochester Castle, having fallen into the hands of the disaffected barons, was laid under siege by King John in 1215. John brought up his siege engines against the castle and pounded relentlessly at the walls, his troops working in relays. But the besieged replied with such effect and caused such execution in the ranks of the royal forces that other methods of attack had to be adopted. The king then employed miners to break through the curtain, and when a breach was made and the troops had entered within the bailey, the garrison, after a valiant fight, retired within the keep. The miners then applied themselves to the keep and broke into that also. Even then those within fought desperately, and the troops, suffering great loss, were com-

[1] *Guillaume-le-Breton.* Prose and verse descriptions.

pelled to retreat again and again. At length, having sustained a siege of nearly three months and been brought to the verge of starvation, the garrison surrendered.[1] The south-east angle of the keep of Rochester castle was rebuilt about this period and is very probably the point where the breach was made.

SIEGE OF DOVER CASTLE, 1216

In the following year and on the invitation of the insurgent barons, Louis the Dauphin of France, with a strong force, crossed the channel and laid siege to Dover castle. With his powerful petrariæ and other siege engines Louis made a violent and incessant attack on the walls. The garrison, under the leadership of its constable, Hugh de Burgh, replied with such devastating effect that the French, feeling their loss, moved both their camp and their engines further back. Meanwhile King John died and Louis and the barons, who were before Dover, thinking that England was now in their power, called upon the constable to surrender, offering him great honours and high position. But Hugh and his fellow knights refused to surrender and the siege was raised.[2]

Among the examples of siege works still existing are those at St. Andrews Castle, Fifeshire, made during a siege in 1546–47. Here the attacking forces drove an underground tunnel towards the castle from a point about 130 ft. from the walls. Countermining was then undertaken by the garrison, and after some tentative efforts to locate the advancing mine they eventually broke into it at a point about midway between the walls and the starting point of the mine. The countermining was so exact that although the mine had deviated considerably from a straight course the countermine broke into it at its end, and at the convenient level immediately above the heads of the enemy. Both mine and countermine are still preserved.

[1] *Roger of Wendover.*
[2] Ibid.

Conway Castle, from the North-East. p. 157.

Conway Castle, from the South-West. p. 157.

Caernarvon Castle from the South-East.   p. 157.

CHAPTER XIII

EDWARDIAN AND CONTEMPORARY CASTLES

THE course of this history has now reached a period, the reign of Edward the First of England, when some of the most powerful castles of any age or country were built in Great Britain. Extensive experience in sieges, both at home and abroad, had shown the king and his barons the weak points of existing fortifications, and in the new castles which they built these defects were rectified. Though considerable attention was still paid to outworks, the general tendency was to concentrate the central defence on a four-square castle, surrounded by one or two lines of walls, and having a strong round tower at each corner of the inner walls. Powerful gateways now take the place of keeps, and there is a more liberal provision of gateways and posterns.

A serious defect in the earlier castles lay in the fact that there was generally only one gateway and one postern. The last word in war, whether carried out on a small or on a large scale, must always be starvation. A king might be secure behind the walls of an impregnable castle, but, unless he had adequate means of bringing in supplies, of making sorties, or of effecting escape, in the event of a siege by a stronger force than his own he was doomed sooner or later to capture. In the new castles more gateways were provided and, in many cases, the difficulties of investment were increased by extensive outworks. The design of the inner bailey also secured greater mobility of defence forces and facility of command.

One of the earliest of these castles was that of Caerphilly, Glamorgan, about 1267–1277. Caerphilly castle stands on what was an island in a lake, the lake being fed by a stream and held in by a great screen wall, or dam, forming the barbican (p. 154). The central, or main, portion of the castle is rectangular, and is surrounded by two lines of walls; the inner wall having a tower at each corner and two large gatehouses, one in the middle of the east wall and another in the west wall. The towers have such bold projection beyond the corners that the outside faces of the walls between them were completely commanded and could be swept by projectiles from their meurtrières and battlements from end to end. The outer wall, which is lower and not so thick as the other, has gateways on the east and west but no corner towers. The east gateways look towards the barbican, from which they were approached by a drawbridge. The west gateways look towards an outwork, which stood in the same lake as the main buildings and was reached from

154   CASTLES

them also by a drawbridge. The hall and living-rooms are built against the curtain on the south side of the inner bailey, and the kitchen, bakehouse, and other domestic offices in the space between the two walls, behind the hall.

In addition to the main gateways there are three posterns or sally ports in the inner bailey, one on the south and two on the north; and two in the outer

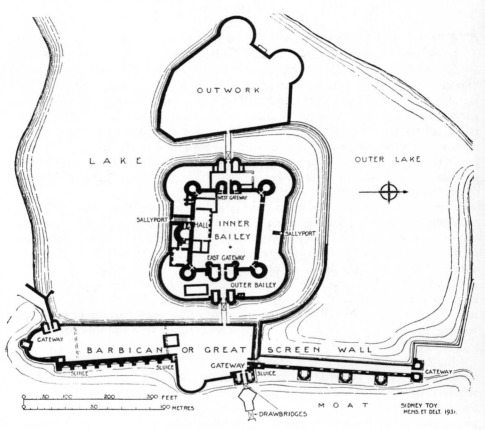

Plan of Caerphilly Castle.

bailey, one on the north and the other on the south. The postern on the south of the inner bailey is in line with that on the south of the outer bailey and opens directly out of the hall to a vaulted stairway down to the doorway in the outer wall. The inner and outer doorways are each protected by a portcullis. From the posterns in the outer wall supplies could be brought in, sorties made, or escape effected by boat across the lake. The circulation of the inner bailey was greatly facilitated by a mural passage, which was carried all round the walls, about 15 ft. above ground level, and checked only by portcullises on either side of the gatehouses.

The massive screen wall, sustained on one wing by a series of huge buttresses and on the other by three strong towers, is a most powerful and imposing work of military engineering (see pp. 154, 124, 211). Through it run three sluices, by which the level of the water in the lake was regulated, and behind it is the long and spacious barbican. The stronghold of Caerphilly was the east gateway of the inner bailey, which could be held as well against an enemy within the bailey as against one without.

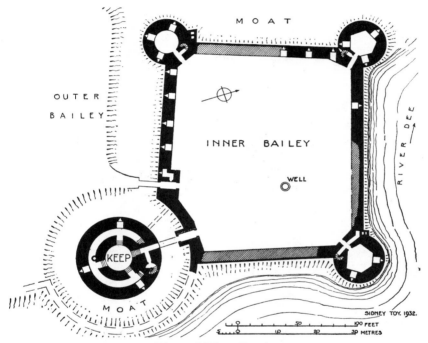

Plan of Flint Castle.

Though the underlying principles of development may be maintained, the design of a castle must always depend largely on the character of the site it occupies and the purpose it has to fulfil. Again the defensive factors of surprise and secrecy themselves demand endless varieties of plan and disposition of parts. Flint castle, built in 1277–80, Conway, 1283–87, and Caernarvon, 1285–1322, each formed part of a general scheme which included a fortified town.

At Flint the ultimate stronghold is a cylindrical keep, which stands isolated at one corner of a rectangular fortress; like the donjon at Lillebonne

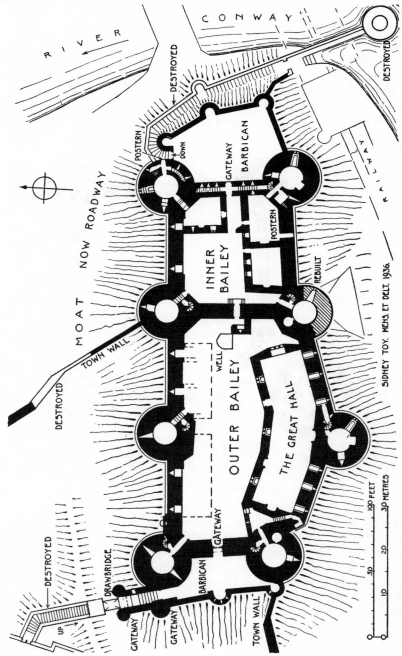

Plan of Conway Castle.

and the Tour de Constance at Aigues Mortes. The keep at Flint, only the lower portion of which remains, was a powerful structure 71 ft. in diameter and probably three storeys in height (p. 155). It was approached from the inner bailey by means of a drawbridge over the moat and entered at a level midway between the basement and the first floor. From the entrance passage steps in front led down to the basement and a wide spiral stairway on the left to the upper floors.

Conway castle stands on a high rock on the shore of the estuary and, following the contour of the rock, is long and relatively narrow. It is defended by eight towers and has a gateway at either end, each gateway being flanked by two of the towers and covered by a barbican. A cross wall divides the castle into two baileys of unequal size. The larger, or outer, bailey contained the great hall and the domestic offices of the garrison; the inner bailey the royal apartments and private offices. The gateways are constructed through the curtain wall and are defended only by the adjoining wall towers; there are no gatehouses. The security of the castle depends largely on the difficulty of access to it. On the west, the town side, entrance was effected only after climbing a steep stairway, passing over a drawbridge and through three fortified gateways; all in face of direct fire from the towers and walls on every side. The approach from the estuary was commanded for the whole of its course by the east barbican, which towered high above it, and by a tower in the estuary itself (pp. 156, 151).

In the inner bailey there are two posterns in addition to the east gateway. One of them is in the south wall and stands high above the rocks on the edge of the river. It would be of service in an emergency, like those at Carcassonne, by the use of a rope ladder. The other is in the north-east tower. This tower contains a beautiful little chapel and has, in addition to the postern, means of exit by stairways and passages in all directions; it probably contained the royal chambers. The circulation of the wall walk is uninterrupted all round the curtain and on the cross wall; thus enabling the defensive forces to be rushed speedily to any desired point. All the towers, on the sides facing towards the field, have beam holes for hoarding, and four of them have high turrets from which the approach or distant operations of an enemy could be observed.[1]

Caernarvon castle stands on relatively level ground, though there is a fall from the east, where there was a mound, to the west. The plan resembles an hour-glass, narrow in the middle and bulbous at both ends. It is a very powerful structure, surrounded by strong and lofty walls and defended by two gatehouses and nine wall towers, many of the towers having turrets as at Conway. The north side of the castle faces towards the town and the south side towards the river. One gatehouse, called the King's Gate, stands in the middle of the north wall and provided direct entrance from the town to the

[1] *Vide* "The Town and Castle of Conway" by the Author. *Archæologia*, Vol. LXXXVI, 1937.

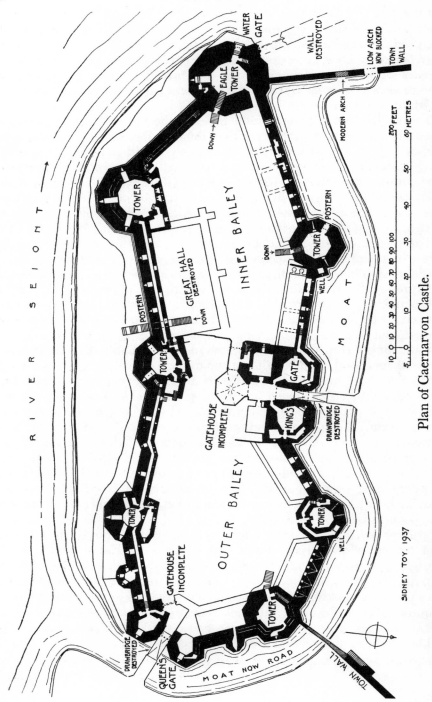

Plan of Caernarvon Castle.

inner bailey in the west portion of the castle. As originally designed this gatehouse was to extend across the castle to the south wall and form the division between the inner and outer baileys. But at present it extends only part way across, and was probably never completed. The Queen's Gate is the entrance from the east, beyond the walls of the town, to the outer bailey. (pp. 158, 152).

In the inner bailey there are three posterns; one from a tower in the north wall, opening on to the moat between the castle and the town; one in the curtain on the south, giving access to the river; and the third, called the Water Gate, giving access to the river from the Eagle Tower at the extreme west of the castle. The Eagle Tower is an especially strong and well defended building, and preserves in its design many of the elements of a keep. From it the whole fortress could be commanded; it contains a private chapel, and has a gate by which entry could be made or escape effected (p. 167).

Attached to the north side of the Eagle Tower are the remains of a wall 13 ft. 6 in. thick, which jutted north-west out into the river. This fragment contains one jamb of a low gateway with the groove for the portcullis by which the gate was closed; and, at a higher level, an entrance passage into the chamber over the gateway. It has been suggested that the space north of the tower and west of the town wall to which this gateway admitted was occupied by a gatehouse. But since the gateway is at water level and there was no through way except by water, it is more probable that the space was occupied by a dock, that the fragment is the remains of a screen wall similar to the "Gunners' Walk" at Beaumaris, and that the gateway was one of two fortified posterns into the dock. The town wall here is not diminished in thickness on approaching the tower, as it was on the east side of the castle and as it is on both sides at Conway.

The curtain and towers on the south front of the castle have two tiers of mural passages with meurtrières to the field. On the north front, towards the town, some of the meurtrières are so constructed that from one recess archers could shoot in two or even three different directions; in others openings from three recesses converge towards one loophole. When put in a state of defence the castle was bristling with archers standing at different levels behind the defences. On the south side, including those on the battlements, there would be three tiers of armed men.

A continuous wall walk, like that at Conway, though it facilitates the operations of the garrison while it is held, is a serious menace when one part of it is carried by assault, since the whole could then be overrun. At Caernarvon all the towers stand astride of the walk, so that if any section of the curtain was carried it could be isolated by closing the passage through the towers.

The internal buildings are now represented principally by foundations, but the towers and curtain walls are practically intact. Caernarvon castle, with its powerful fortifications almost complete, ranks among the finest examples of mediæval military art.

Harlech castle, Merioneth, built 1285–1290, and Beaumaris castle, Anglesey, built between 1295–1320, both show the development of the principles of defence suitable to the sites they occupy.

Harlech castle stands on a hill. (See also p. 138). It consists of a rectangular fortress enclosed by two lines of walls, forming the inner and middle baileys; and an outer bailey, which extends down the precipitous slopes of the

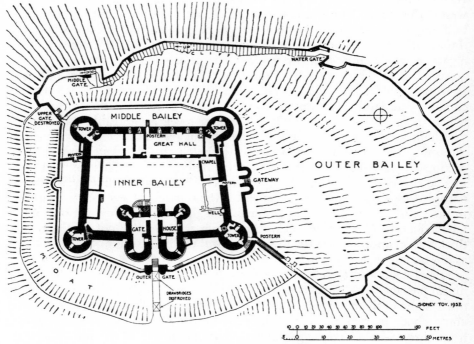

Plan of Harlech Castle.

hill and covers only the north and west sides of the main fortress. The east and south sides are defended by a wide moat. The inner bailey has a tower at each corner and a large gatehouse in the middle of the east front, the towers having the same bold projection as those at Caerphilly. There is a postern in the north wall and another, leading out of the great hall, on the south. The middle bailey, actually a narrow terrace between the two walls, is obstructed at one point by a cross wall with a doorway. It has two gateways, one in front of the gatehouse and the other, on the north, opening to the outer bailey, and two posterns, both opening to the outer bailey.

EDWARDIAN AND CONTEMPORARY CASTLES

The outer bailey presents to the field a rising series of precipitous rocks; and the approach through it, from the Water Gate at the front of the hill, to the upper fortress was by way of a steep path, cut into the cliff face along the west curtain. Any other line of approach would involve a perilous climb over rocks to the north gateway. The road up by the path was intercepted by a drawbridge and gate and ended in another gate, below and commanded by the south-west tower. A wing wall, jutting out from the main fortress, intercepted the passage from one part of the outer bailey to the other. From these sides, therefore, the north and west, the castle was strongly defended both by nature and by military art.

But the vulnerable side of the castle is the east front, which faces towards higher ground; and it is here that the military strength of the fortress is particularly concentrated. The east front of the inner bailey is powerfully built. It has a gatehouse in the middle and a strong tower at either flank; one panel of the wall is 12 ft. 4 in. thick and the other 9 ft. 8 in. As at first proposed the other three walls were to be only about 6 ft. thick, like those at Criccieth castle (p. 138), but after they had been carried up to about one-third of their present height it was decided to make them stronger, and as finished they are about 9 ft. thick. The approach from the east was by a bridge with a drawbridge at either end; the passage across the bridge being under attack not only from the gateways, walls, and towers in front but also from the wide wall walk of the outer bailey on the right flank. Beyond the bridges the way to the inner bailey was obstructed by the outer and inner gateways. The outer gate was closed by a two-leaved door only; but the long passage through the inner gate was closed by a stout timber bar, three portcullises, and two doors, while it was commanded from above by eight wide machicolations.

At Harlech the gatehouse was the stronghold of the castle, and before the erection of the later stairway in the courtyard, could be held against the inner bailey as well as against the exterior; the inner system of door, portcullis and machicolation being reversed against the bailey. The gatehouse stands astride of the wall walk of the curtain and commands it by doorways on both sides, but otherwise the walk is continuous all round the walls.

Beaumaris castle stands on level ground on the sea shore, and the disposition of the defences is much more regular here than at Harlech. It consists of two lines of fortifications, forming the inner and outer baileys, and, except at one point on the south where the sea enters to form a small dock, is surrounded by a moat (pp. 162, 137). The inner bailey is almost square, its walls are about 15 ft. 6 in. thick and are defended by six towers and two large gatehouses, the gatehouses being opposite to each other like those at Caerphilly. Each of the gateways was defended by three portcullises and two doors, and, as far as the remains indicate, by machicolations; all arranged in a manner similar to those of the inner gateway at Harlech, and either of the

162   CASTLES

gateways could be held as well against the inner as the outer bailey. The south gateway was further protected by a small barbican. Both the gatehouses were substantial structures containing large halls and rooms. In the intermediate tower on the east there is a beautiful vaulted chapel, practically intact.

The wall walk is continuous except at the gatehouses, where it is barred by

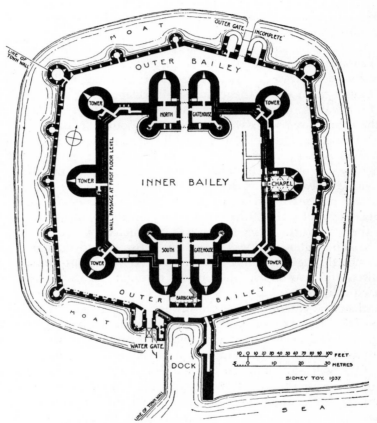

Plan of Beaumaris Castle.

doorways on both sides. The doorway on either side of the north gatehouse is placed slightly clear of the inner face of the wall, leaving an open space across the angle between the walk and the doorway. This space was spanned by a bridge which was probably of timber and could be removed as desired, thereby completely isolating the gatehouse from the walk. Vestiges of the stone supports of the bridges remain. This gatehouse was the stronghold of

EDWARDIAN AND CONTEMPORARY CASTLES

the castle. Circulation of the castle was greatly facilitated by mural passages, at first-floor level, which run all round the walls, except for a short portion on the north-west, and into the gatehouses.

Where suitable stone slabs were difficult to obtain the surfaces of wall walks were frequently covered with lead, and the grooves for the lead flashing are still to be seen in the parapets of many castles, as at Conway. At Beaumaris a considerable portion of the lead flashing itself remains in the parapets at the north-east corner of the walk.

As finished about 1290 the inner bailey formed the whole castle, the outer wall was not built until about 1316–20. This wall is constructed in nine panels, is strengthened by many wall towers, and has gateways on the north and south. The gateways are placed considerably out of line with the inner gatehouses, thus involving the exposure of the flank of an enemy entering the outer and proceeding to the inner gates. The north gateway is in a fragmentary condition and, probably, was never completed. Near the south gateway there was a small dock, running in from the sea between the town walls and a great screen wall, called the Gunners' Walk, which juts out from the outer wall of the castle. A door at the head of the dock gave admittance into the outer bailey. The screen wall defended the dock on all sides, as well from a mural passage and chamber within as from its walk and battlements above.

At Villandraut, thirty miles south of Bordeaux, there is a rectangular castle, built 1306–7, which is of similar character to those in Wales, but there is only one curtain wall and one gatehouse. This castle has a tower of bold projection at each corner, a gatehouse in the middle of the south wall and a postern in the north wall. Here the towers are vaulted in each storey, and in order to facilitate the construction of the vaults they are hexagonal internally though round on the outside. The castle is surrounded by a moat.

TRIANGULAR AND CIRCULAR CASTLES

Caerlaverock Castle, Dumfries, following the contour of the rock on which it stands, is constructed on a triangular plan. The first castle probably dated from the latter years of the thirteenth century. This building was destroyed in 1312, afterwards rebuilt and again destroyed in 1356. It was rebuilt a second time at the end of the fourteenth or beginning of the fifteenth century. But at each reconstruction the original plan of the castle as described in a poem of about 1300 has been preserved and probably much of the walling incorporated. The contour strongly indicates that this, and not what is called the "old castle" a short distance from it, is the original site. There is a large gatehouse, flanked by towers, at one angle (p. 168) and a circular tower at each of the other two angles. The castle is surrounded by a moat.

The Castello di Sarzanello, forty miles north of Pisa, built about 1325, is also designed on a triangular plan. Here there is a round bastion, open at the back, at each corner; the bastions being continuous with the walls, and of the same height. The walls are high, are strongly built, and, with widespread plinths rising half-way up their sides, present a very formidable

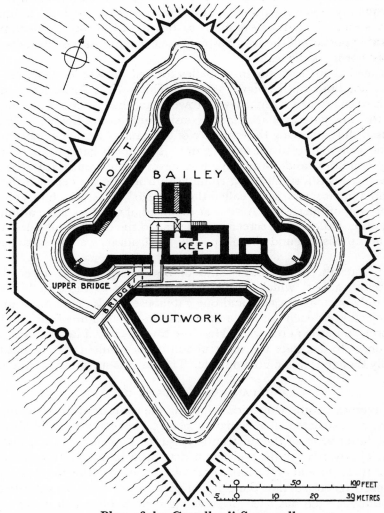

Plan of the Castello di Sarzanello.
After Bodo Ebhardt.

appearance; they are defended all round by machicolations at the level of the battlements. The gateway enters through the south wall under the protection of a rectangular keep, which is built against the inside face of the wall and overlooks the approaches as well as the bailey. The castle is defended on this side by an outwork, which is also triangular and stands base to base with the other, a moat passing between them.

The outwork has no bastions, but its walls are of the same height and design as those of the main building and their battlements are connected by a high bridge across the moat. An encircling moat is carried round outside all the walls and there is a wall beyond the moat. Entrance to the castle is by way of a long bridge, which, after crossing the encircling moat and passing along by the side of the outwork enters the cross moat below the high bridge. Then, after making two turns, it ends in the entrance gateway; and during its

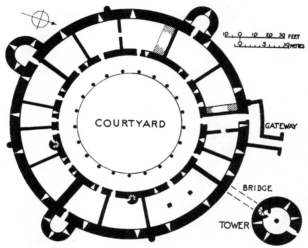

Plan of the Castillo de Bellver.
After Lamperez and Romea.

whole course it was under direct attack from both the outwork and the main building.

The Castillo de Bellver, Mallorca, and the Château de Montaner, Hautes-Pyrénées, the first dating from the early part of the fourteenth century and the second from the latter part, are both built on a circular plan; and Queenborough Castle, Sheppey, Kent, built 1361–77 and destroyed in the seventeenth century, was also circular. In all three the halls and living-rooms were built round inside, between the curtain and an inner wall concentric with it; much in the same manner as in the shell keeps of an earlier period.

At Bellver there is a semi-circular wall tower on each of the east, west, and south sides of the curtain. On the north side there is a strong round tower, which stands clear of the castle but for a high single span bridge between the tower and the battlements of the curtain. The gateway and its elbow shaped approach is placed near this tower and is completely commanded by it.

The Château de Montaner is built of brickwork. Here the curtain is supported all round by buttresses and interrupted at one point by a square tower, the donjon of the castle. The entrance gateway is on the side of the castle opposite to the donjon. The château was surrounded by a moat.

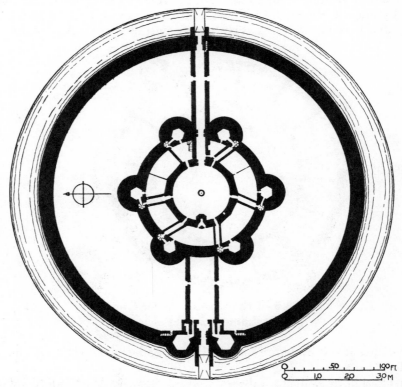

Plan of Queenborough Castle. From Hatfield MSS.

Queenborough differed from the other two castles in being enclosed all round by an outer bailey and a particularly strong outer wall. This outer defence was concentric with the inner curtain, was surrounded by a moat, and has a gateway on the west side and a postern on the east. Each of the passages across the bailey from the outer gates to the inner curtain had a screen wall on either side to prevent the bailey being overrun in the event of the outer defences being carried by assault. The inner curtain was much higher than the outer wall and was strengthened by six round wall towers,

Caernarvon Castle. The Eagle Tower from within the Bailey.

Caerlaverock Castle. The Gatehouse. p. 163.

two of them being placed close together in order to defend the gateway which passed between them.

Among the finest military works of the fourteenth century are the fortifications of the city of Rhodes, rebuilt by the Knights of St. John after their acquisition of the island in 1310. The walls and towers surrounding the city are of great strength and height and have strongly fortified gateways. The land walls are particularly powerful. For the purpose of defence they were divided into several sections, each section being assigned to a contingent formed of those knights speaking the same language; of the sections, that allotted to those speaking the English tongue is especially strong and imposing. Here the walls are defended by two tiers of battlements, the lower tier formed by a terrace which runs along from tower to tower in front of the face of the wall, and the upper on the top of the wall itself, which rises to a great height behind the terrace.

CHAPTER XIV

TOWNS, FORTIFIED BRIDGES, AND TOWERS

FROM the earliest times cities grew up below or were associated with a citadel which occupied a dominating position on one side of their defences. A Norman town frequently stretched away from the foot of the castle of the lord to whom it belonged, the mound with its keep having the bailey on one side, towards the field, and the town on the other, the mound often jutting well into the town. The castle stood in a commanding position from which the garrison could either protect the town, or defend themselves against it in the event of its fall or of the disaffection of the townsmen. Totnes, Launceston, Pleshey, and Gisors are examples of this plan.

In the latter part of the thirteenth century many new fortified towns were built in the south of France and in England and Wales; those in Britain as well as a large number of those in France being founded by Edward I, King of England and Duke of Aquitaine. Where the site permitted these towns were rectangular, as Monpazier in France and Flint in Wales, but many are irregular in shape, as Libourne, near Bordeaux, and Winchelsea, near Hastings. There were always parallel streets running through the town from end to end and others cutting them at right angles and dividing the town into rectangular spaces. One of the spaces was the market place, which had therefore continuous streets passing along on all four sides but none crossing the centre. The church stood in a square generally near the market place.

In laying out the plan of the defences of a town some especial conditions had to be taken into consideration. Unlike a castle, the space within the walls was occupied by blocks of houses, which, if not properly planned, would greatly impede the circulation of troops. The open market place, forming a rallying point in time of siege, was generally placed near the centre of the town, and from it streets led towards the gateways and the walls. As in Roman camps and towns there was a road, called the *pomerium*, all round the defences behind the walls. This road gave direct access to the curtain and its towers at all points and enabled the rapid movement of troops and engines from one part of the fortifications to another.

The towers of town walls were often built with no wall at the back and show towards the town an open gorge, either crossed by a stone arch at the level of the wall walk on the curtain, or quite open from base to summit save for a timber floor at that level. They project principally on the outside and on the inside their lateral walls after passing across the wall walk end abruptly either flush with the inside face of the curtain or slightly beyond it.

Towers of this character are seen in the fortifications of Visby, Conway, Caernarvon and Avignon, the first three dating from the thirteenth century, and the last built 1350–1374. They are, however, by no means characteristic of the defences of towns but occur frequently in the curtains of castles, as at Framlingham, about 1200, and Corfe, about 1270, in England; at Este, Villafranca, and Verona, in Italy, all three dating from the fourteenth century; at Smederevo, on the Danube, built 1432; and at Roumeli Hissar, on the Bosporus, in walls built in 1452.

Often the tower was entered from the walk by a doorway on one side only, and to pass from one side of the tower to the other it was necessary to ascend to its battlements; so that if one section of the curtain was taken by the enemy that section could be isolated by removing the timber stairways in the tower at either end. This arrangement can be seen clearly in the town walls at Conway though the work is in places broken away. On one side of the tower there was a doorway from the wall walk to the timber floor of the interior. But there was no doorway on the other side. To reach the walk on that side it was necessary to ascend to the battlements by a wood stairway within the tower and descend by a flight of stone steps built against the tower wall on the outside.

But these open-backed towers were relatively weak, and the chief reason for their employment was probably a saving of material and labour in construction. Once an enemy had penetrated within the town or castle they could no longer be held, and not all of them could be used as a check to the circulation of the wall walk like those at Conway. At Avignon there are doorways to the wall walk on either side of the towers and a stone bridge across the gorge from one doorway to the other. At strategic points in the curtain, such as sharp angles and places particularly liable to attack, towers, complete all round, were usually built. At Conway the tower at the west angle of the town is closed at the back to the height of the wall walk of the curtain though open above that level, while that at the north angle is closed to its full height.

In order to protect the curtain between the wall towers, smaller towers, or turrets were often built, either rising from the ground or corbelled out from the wall. At Visby, Gotland, midway between each pair of large wall towers there is a smaller tower, which is corbelled out from the wall on both faces and sits upon the curtain like a saddle (p. 172). These towers are also open at the back. At Avignon there are, in places, two turrets between each pair of wall towers. Here the turrets are flush with the inside face of the wall, but on the outside rest on buttresses which rise from the ground; their battlements are reached by flights of steps up from the wall walk of the curtain on either side.

The walks on the curtain walls of towns, as well as of many castles, are gained at many points along the fortifications by means of flights of stone steps, built against the inside face of the walls.

The gateways of a town were more numerous than those of a castle and their positions in the curtain were largely governed by the situations of the highways on which they opened. If one of the approaches to the town was by way of a bridge across a river it was important that the bridge should be fortified and protected on the far side by a barbican or *Tête-de-pont*.

Many fortified bridges of the Middle Ages still exist throughout Europe and among them that at Tournai, built in the thirteenth century, the Puente de Alcántara at Toledo, mainly of the same period, and the bridges at Verona,

Visby. The Town Walls. pp. 171, 203.

Orthez, and Cahors, all dating from the fourteenth century, are particularly fine and well-preserved examples.

At Tournai the Schelde runs through the middle of the town, and at the two points where it passes through the defences the curtain walls on either side were connected by fortified bridges. That at the north side of the town, the Pont des Trous, still exists. It is a covered bridge of three spans, pierced by loopholes on both sides and defended by a strong square tower at either end. The Ponte Castel Vecchio, Verona, was built by Can Grande II as part of the fortifications of his castle and was defended by the great square tower of the castle on the town side of the bridge, and originally also by a tower and two drawbridges at the far side.

The Pont Vieux at Orthez, Basses-Pyrénées, crosses the river Gave in three unequal spans. It is defended by a tall tower, which stands across the road over a pier at the middle of the bridge, and is pierced by a fortified gateway. The Pont Valentré at Cahors, Lot, one of the most imposing bridges of mediæval times, is of six spans and is defended by a tower at each end in

TOWNS, FORTIFIED BRIDGES, AND TOWERS

addition to a third in the middle. All three towers stand across the road, each of them is pierced by a fortified gateway and defended by an embattled parapet. The two end towers have also machicolations, placed immediately below the parapets. This bridge was defended at the end opposite the town by a *Tête-de-pont*.

One of the best examples in Britain is the bridge over the Monnow at Monmouth, built about 1272 and spanned by a fortified gatehouse about 1296. Here the passage through the gateway was defended by machicolations, a portcullis, and a two-leaved door.

Pont Valentré. Cahors. p. 172.

TOWERS

During the fourteenth century some of the existing castles were strengthened by the addition of powerful towers, built either within the bailey, as at Foix, Ariège, or in the curtain walls, as at Warwick.

The château de Foix stands on the summit of a precipitous hill and consists of two baileys, one within the other, defended by a barbican on the east and an outwork on the west. In the inner bailey there are three towers (pp. 174, 175).

The nucleus of the castle is the rectangular tower which stands at the north end of the bailey. It dates principally from the eleventh century but incorporates masonry of a still earlier period, composed of courses of rubble with lacing courses of Roman brick. The machicolated parapet was built probably in the fifteenth century. To this tower a long hall, one storey in height, and a second rectangular tower, at the other end of the hall, were added at a period now difficult to determine, owing to later and modern alterations. The hall and both towers are shown on a seal of 1215, and it is probable that the additions were made in the twelfth century.

About the middle of the fourteenth century the second tower was either remodelled or completely rebuilt and as then finished was the donjon or keep of the castle; being much larger and stronger than the north tower and containing the principal living-rooms. It is three storeys in height, is vaulted in every storey and has a flat roof with machicolated parapet. The second storey is a lofty apartment with one window and a small mural chamber, doubtless originally a latrine, but no fireplace. The third storey,

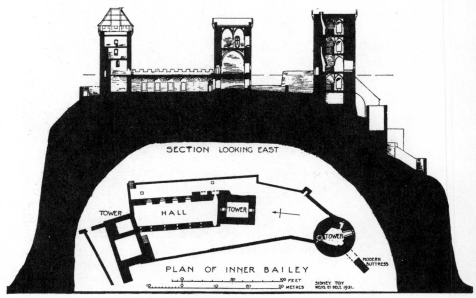

Château de Foix.

which is vaulted in two bays, has two large windows, a fireplace, and a mural chamber similar to that below. At the south end of the west side of the room there is a small window, reached by a flight of steps up; and, beside the window, an opening and flight of steps down lead straight through the wall to a postern on the outer face of the tower. This postern provided a way of escape by means of a rope ladder down to the building, ruins of which, incorporated in later work, still exist on this side of the tower.

The entrance doorway of the tower is on the second storey, opening on to the flat roof of the hall; and from the entrance passage a spiral stairway leads to the upper floor and the battlements. Owing to the fact that some of the early work has been destroyed and that alterations and repairs have been effected, it is not clear how the roof of the hall was approached from the

Château de Foix from the West.

Craigmillar Castle from the South-East.   p. 180.

ground. The existing flight of steps, built outside against the hall and blocking one of the hall windows, and the doorway cut into the tower stairway at the head of the flight, are modern. But whatever may have been the way up, once gained and the approach cut off, this flat roof formed a magnificent fighting platform, defended by battlements on either side and a strong tower at either end. The round tower at the south of the bailey was added in the fifteenth century.

Warwick Castle consists of an oblong bailey, running east and west, and a high mound at the west end of the bailey. It is defended on the south by the

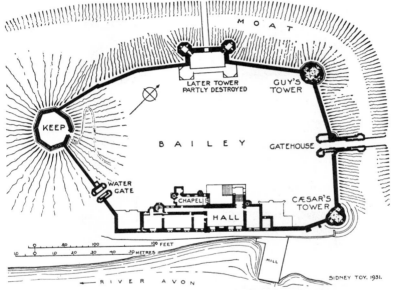

Plan of Warwick Castle.

river Avon and on all other sides by a wide and deep moat. On the summit of the mound are the fragmentary remains of a polygonal shell keep, dating probably from the twelfth century. The keep stood partly within but mainly outside the bailey. The existing fragment consists of the three sides that were within and is joined to the curtain wall at either end. To this fragment battlements, built flush with the inside face of the keep in continuation of those on the curtain, and two turrets have been added. These additions were made, probably, in the early years of the seventeenth century. (See also p. 64.)

The castle suffered considerably during the Civil Wars of Henry III. But it was raised to a fortress of great strength in the fourteenth century when two powerful towers were added, one at the north-east and the other at the

178 CASTLES

south-east of the curtain, and a gatehouse was built in the middle of the east wall. Extensive living and service quarters were also built against the curtain overlooking the river. The tower at the south-east is called Cæsar's Tower, and the other Guy's Tower; both were built during the latter half of the fourteenth century.

Cæsar's tower is three-lobed on plan and rises to the height of 133 ft. from the foot of the massive plinth outside, 49 ft. below the bailey, to the top of the

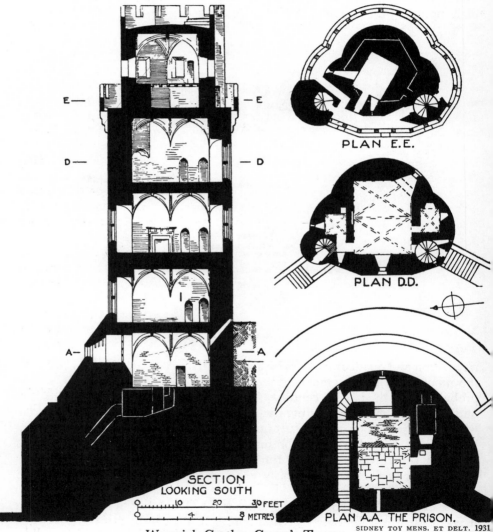

Warwick Castle. Cæsar's Tower.

upper parapet (pp. 178, 199). It consists of six storeys; all, except a small rectangular chamber forming the fifth storey, being covered with thick stone vaulting. The lowest storey is below the level of the bailey, from which it is entered directly by a doorway and a flight of steps down. The next three storeys are alike in design. In each there is a large rectangular room with a fireplace, a mural chamber and a latrine. There are two tiers of battlements, and the last two storeys of the tower, the uppermost of which is a large hexagonal hall for the guard, rise behind a narrow gallery at the level of the first tier. The lower parapet is machicolated all round, but the upper one is crenellated only; in both parapets the merlons are pierced by cross-shaped loopholes.

The approach to the tower was from the walk on the south curtain wall leading from the domestic buildings, and the entrance doorway stands at the head of a flight of steps up from the walk. This doorway admits directly to the spiral stairway at a level about 3 ft. below the floor of the fourth storey; and the stairway has doorways to each of the floors, except the fifth, which is entered from the lower battlements. The doorway opening into the bailey at the foot of the stairs is modern. On the north side of the tower a second stairway descends from the lower battlements to the north curtain. This stairway has now been broken into on the inside to form a passage through from the fourth storey. But originally, to pass from one section of the curtain to the other, it was necessary to mount the stairway on one side of the tower, traverse the gallery at the lower battlements, and descend the stairway on the other side.

Guy's Tower is polygonal in plan; it is five storeys in height, and is surmounted by a single line of machicolated battlements. The arrangements of the internal chambers and stairways are similar to those of Cæsar's Tower. All the storeys are vaulted.

In Scotland and in the border counties of the north of England, fortified towers, with no other outer defences than the wall enclosing the courtyard, called the barmkyn, in which the tower stood, were now being built in large numbers. There are early examples of such towers, as at Pendragon in Westmorland, built in the twelfth century, but they are principally of the fourteenth century and later periods. These towers, though varying in size and internal arrangements according with the status and means of their builders, are generally rectangular in plan, are three or four storeys in height, and have thick walls crowned by embattled parapets. The upper storeys are reached by spiral stairways, formed at the corners of the towers and terminating in turrets, or cap-houses. Either the first or the second storey is covered by a barrel vault; when it is the second, that storey is formed by building a floor in line with the stringing of the vault. Often one or two of the other storeys are also vaulted, and when the uppermost storey is vaulted the roof consists of flat stone slabs laid directly upon the vault.

The entrance is usually either on the first or on the second storey. When on the second storey it was reached by a movable stairway, and there was often also an outer doorway to the first storey, as at Closeburn, Dumfriesshire. The first storey formed the stores and sometimes, in addition, contained a small prison. The upper storeys contained the hall, the great chamber and the living-rooms. Mural chambers were formed at the corners of the tower and in some cases, as at Craigmillar and Chipchase, greater space was obtained at one corner by a projecting wing which was carried up the full height of the tower and contained the entrance lobby at ground level and mural chambers in the upper storeys. Clearly the privacy and social amenities of such confined quarters were restricted and additional accommodation was often obtained by subdividing the storeys. Chipchase, Northumberland, dating from about 1350, Craigmillar, Midlothian, and Threave, Kirkcudbright, both built in the latter part of the fourteenth century, are all good examples. Many of the towers were subsequently incorporated in other defences, as Craigmillar, where the present inner bailey, with buildings round the courtyard, was added in the fifteenth century, and the outer bailey in the sixteenth century.

Chipchase Tower stands on the left bank of the North Tyne, ten miles north of Hexham. It is built of stone in regular courses and measures externally 51 ft. 6 in. by 34 ft. It is four storeys in height; the first storey, at ground level, being covered by a barrel vault and the other storeys originally by timber work. On the south-east corner is a projection, 3 ft. deep by 20 ft. 9 in. wide, running the whole height of the tower and containing the entrance doorway, at ground-floor level, and providing space for mural chambers at the other floors. A spiral stairway, beginning at the entrance passage and rising to the battlements, is built at this point. Turrets, rounded on the outside, rise at the corners of the tower from corbels at the level of the battlements; and both the main building and the turrets were defended by machicolated parapets (p. 183).

The entrance doorway was protected by a stout oak portcullis, and a door secured by two timber bars; the portcullis, which still remains in position, being operated from a small chamber over the doorway. The first storey was the store-room, and has neither windows nor ventilating shaft. At one end there is a well; the water being drawn, apparently, from the second storey through a trap-door in the vault.

The second storey, containing a fireplace and a double locker, was but dimly lighted by two loopholes. The third and fourth storeys were both well-lighted halls and were provided with fireplaces, latrines and mural chambers. An oratory with stone altar table and piscina is formed at the north-east corner of the third storey, and the kitchen at the south-east corner of the fourth storey.

Craigmillar Castle is built on an eminence, and its great tower, among

TOWNS, FORTIFIED BRIDGES, AND TOWERS 181

the finest and best preserved of this type, stands on the edge of a low cliff, the baileys stretching out on either side of it, and at the back. The tower consists of an oblong portion and a small wing, the wing being on the side towards the cliff. The main portion is of four storeys. The second and fourth storeys are vaulted, and each of these storeys is divided from that below by a timber floor, built in line with the stringing of its vault. In the wing the second, third, and fourth storeys are vaulted. The existing fifth storey was added in the sixteenth century. (See also p. 176).

The entrance was most skilfully defended. It is placed in the re-entering

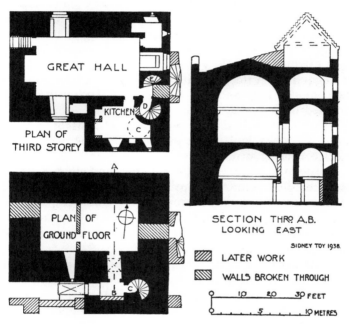

Craigmillar Castle. The Tower.

angle of the wing, and was gained only after passing round at least two sides of the tower and, originally, along the face of the cliff. Immediately in front of the doorway there was a deep chasm, spanned by a drawbridge. The entrance lobby had its floor 3 ft. below the sill of the doorway, and was commanded from above by a small guard-room; so that an enemy who had broken through the door would stumble or have his attention diverted by the sudden drop and be at the mercy of those in the guard-room above. The chasm is now filled in, and the floor of the lobby has been raised. From the lobby a doorway on the left leads, by a lofty passage through the wall, to two doors, one above the other; the lower door giving entrance to the first storey rooms and the upper door to the second storey. The upper door, which must have been reached by a ladder, was therefore so placed that it could be used to defend the passage; it is now blocked. Both the lower storeys were dimly lighted by loopholes.

To gain the upper rooms it was necessary to ascend in succession three spiral stairways, marked C, D and E on plans. The first from the lobby to the guard-room, the second from this level to that of the great hall, and the third from the great hall to the top floor and battlements; the short passage from one stairway to the other, at each level, being blocked by a door. The great hall is on the third storey. It is well lighted by three windows and has a wide fireplace at the west end. On the north side of the fireplace a passage and flight of steps lead down through the wall to a doorway on the outside face of the tower. This doorway, later adjusted to rooms built here and now destroyed, was probably a postern by which escape could be effected. The passage goes straight through the wall and is not likely to have led to a latrine, as has been suggested. A doorway at the north-east corner of the hall opens to a mural chamber and another doorway, at the south-east, leads into a kitchen formed in the wing.

The roof of the tower is of low pitch and consists of stone slabs, laid directly upon the upper vault. The parapet rises flush with the wall faces, without either corbels or string course.

Threave Castle stands on the edge of an island on the river Dee and was defended on those sides not washed by the river by a ditch and an outwork. It consists of a great tower, built by Archibald, Earl of Douglas, known as Archibald the Grim, in the fourteenth century, and an enclosing courtyard the wall of which, now in fragments, was built probably towards the end of the fifteenth century (pp. 184, 185).

The tower measures 45 ft. 6 in. by 24 ft. internally at the base, its walls are 8 ft. thick, and it is five storeys in height; one storey only, the second, is vaulted. The second storey, with a wide fireplace at one end, was the kitchen. The entrance doorway is at this level and from one corner of the room a spiral stairway rises to the upper floors and battlements. The only means of access to the ground storey was through a trap door in the floor of the kitchen. The ground storey has in one corner a large well and in the opposite corner a small vaulted prison or "pit". Near the well there is a stone sink with a drain to the outside. The pit is screened off by walls 4 ft. thick; it has a ventilating shaft and in one corner a latrine, but there are no windows and the only entry was through a trap-door in the vault, opening from the floor of the kitchen.

The third and fourth storeys are well lighted living rooms with fireplaces and latrines; there is also a latrine at the battlements. From the third storey a postern led straight through the wall to a point immediately above the entrance to the tower. When the outer wall was built this postern, apparently, gave access, by a timber bridge, to the chamber over the gateway.

Chipchase Castle. The Tower from the South-West. p. 180.

Tantallon Castle from the South-West. p. 213.

Threave Castle from the South-East.   p. 182.

Two corbels, which may have helped to support such a bridge, have been cut flush with the wall below the doorway. A corbel projecting out from the battlements high above this point is probably the remaining one of two which

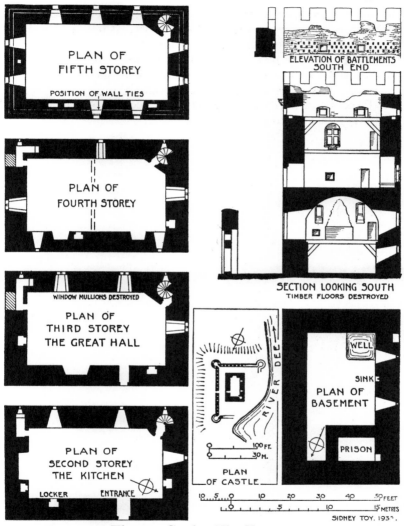

Threave Castle. The Tower.

supported hoarding for the defence both of the postern and the entrance below. At fifth floor level there is a large hole carried through the thickness of the wall all round the tower. It is probably for heavy bonding timbers, originally buried in the walls and long since decayed. The clear span from side to side at this height is 27 ft., and without such powerful ties the lateral thrusts of the roof and upper floors would seriously menace the stability of the walls.

On the face of three sides of the tower and in line with the square openings of the low fifth storey there is a triple row of holes. It has as been suggested that the holes were made for hoarding, but this is very questionable. They are not very deep, conform to no known method of construction of hoarding, and do not occur on the entrance side of the tower where such defence would be mostly needed. The holes are arranged alternately in the same manner as those in a dovecot and were probably made for a colony of pigeons, like those at the summit of the keep at Conisborough castle.[1]

[1] *Vide* p. 98.

Parthenay. Porte St. Jacques. p. 190.

Caerphilly Castle. The East Gateway of the Inner Bailey from within the Bailey. p. 192.

CHAPTER XV

GATEHOUSES AND DEFENCES OF THE CURTAIN DURING THE THIRTEENTH
AND FOURTEENTH CENTURIES

GATEHOUSES

CONSIDERABLE advance was made during the first half of the thirteenth century in the design and construction of the principal gates. The gateways were flanked by towers which formed part of a gatehouse three or four storeys in height, and the approaches were defended from the battlements of the gatehouse as well as from the meurtrières in its towers. The gateway passages were defended by portcullises, machicolations, meurtrières, and two-leaved doors.

Portcullises were generally made of oak, plated and shod with iron, and moved up and down in stone grooves. They were usually operated from a chamber over the gateway by means of ropes or chains, and pulleys, and sometimes also by a winding drum. Machicolations, often spanning the whole passage, opened out in the vault or roof of the gateway. The order in which these defences were arranged varied slightly, but the portcullis was placed in front of the door, which it defended; and the machicolations, either between the portcullis and the door, as at Corfe and Pembroke, or in front of the portcullis, as at Parthenay and the Porte Narbonnaise, Carcassonne. Often there were two or three systems of these defences arranged at intervals through the gateway, as well as an additional machicolation at the entrance.

The gateway into the middle bailey at Corfe, built about 1240, was defended at the entrance by a portcullis, a machicolation, spanning the full width of the passage in four sections, and a two-leaved door. On the inside opening there are vestiges of a second and similar system (p. 190).

At Pembroke the gatehouse of the outer bailey, built about 1250, was defended by a barbican; and in the gateway passage there are two systems of barriers, each consisting of a portcullis, a machicolation, and a door, while beyond the inner door there is a third machicolation. All the machicolations are wide openings in the vault spanning the passage from side to side (p. 191). The passage was further under attack from loopholes in the walls on either side. Projecting from the inner wall of the gatehouse, immediately above the passage, there was a fighting platform or gallery. This gallery was constructed between two stair turrets, which flanked the inner openings. It was entered by a doorway in the north wall of the gatehouse on one side and was protected on the other by an embattled wall, built on an arch thrown across the space between the turrets. The gallery provides for the contingency of the gateway being carried by assault, as, from its commanding position, an enemy rushing into the bailey could be vigorously attacked from the rear.

The gateways at Parthenay, Deux Sèvres, dating from the thirteenth

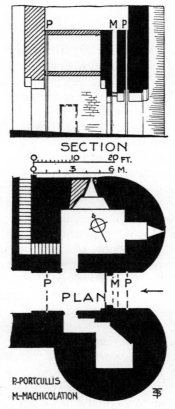

Corfe Castle.
The Middle Gateway. p. 189.

century, are flanked by towers built in prow-shaped form. One of them, leading from the town to the castle, has lost its battlements. The other, the Porte St.-Jacques at one of the entrances into the town, is complete (p. 187). This gateway is defended by machicolations at the battlements of the gatehouse as well as by a machicolation, portcullis and door in the passage.

GATEHOUSES AND DEFENCES OF THE CURTAIN

But gateways reached their fullest and highest development in the second half of the thirteenth century. Some gateways of this period have been noted in the descriptions of the castles to which they belong; it remains to describe some other outstanding examples. The main gateway is often itself defended by an outwork or barbican with its own gate or gates. One of the finest barbicans is that formed by the screen wall at Caerphilly; it

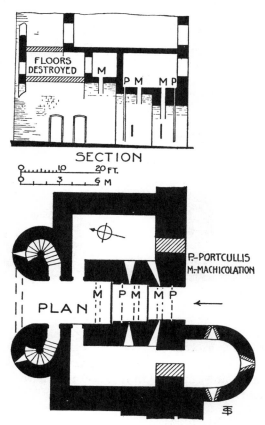

Pembroke Castle.
The Gateway. p 189.

has a gateway in the middle and one at either end. The middle gateway is further protected by a causeway and two drawbridges (pp. 192, 154). When the castle was put in state of defence the outer bridge was probably drawn back over the causeway. But the inner bridge moved on a pivot and had a counterbalance which, when the bridge was raised, descended into a pit, the outer part of the bridge blocking the gateway. The counterbalance greatly facilitated the raising operation, while the pit formed an additional obstacle to those endeavouring to force an entry. The passage was further barred by a portcullis and a heavy door, and was commanded from above by seven machicolations, all stretching across the full width of the gateway.

As already stated the east gatehouse of the inner bailey at Caerphilly was the stronghold of the castle. The outer wall and outer part of the passage have been destroyed. The defence of the inner portion of the gateway was clearly designed against possible attacks from the inner bailey; the doors closed against the bailey and the portcullis was on the bailey side of the doors. In the vault are six square holes, three in front of the

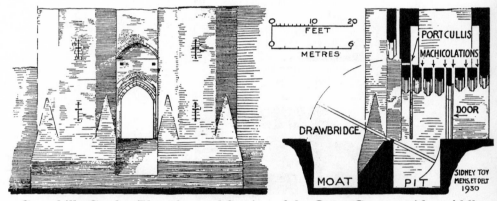

Caerphilly Castle. Elevation and Section of the Outer Gateway (the middle gateway in the Barbican).

portcullis and three between the portcullis and the door. These holes were perhaps for throwing lethal substances, which would spread in falling, on the heads of assailants, or they may have been additional water shoots; there is insufficient height in the recess above for the use of obstructive poles (pp. 193, 188).

A common method of attack on a gate was to pile up faggots or other combustible material against it and set fire to the pile, with the object of burning down the doors.[1] In this gateway provision is made to quench such a fire by pouring water on it from a shoot immediately above the entrance. The outside dressings of the shoot at Caerphilly have been mutilated. But in the gateway of Leybourn Castle, Kent, dating about 1300, there is a similar shoot, which is intact (p. 194). Here the slot, resembling a letter box, measures 1 ft. 7½ in. by 2 in.; it opens out funnel-

[1] *Vegetius*, Book IV, C. 4.

GATEHOUSES AND DEFENCES OF THE CURTAIN 193

shaped in the sill of the window above, and there are indications that it was lined with lead.

The gatehouse of Denbigh Castle, built during the latter years of the

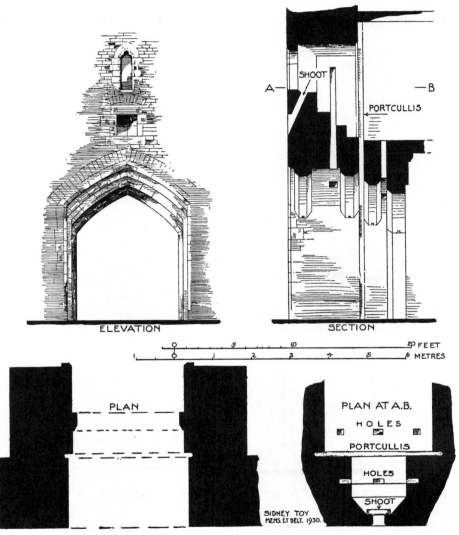

Caerphilly. Details of the East Gatehouse. p. 192.

thirteenth century and now in ruins, was a powerful and skilfully designed structure. It consisted of three towers ranged in triangular form round a central hall; two of the towers flanking the gateway on the north and the third standing within the bailey on the south. There was a moat in front of the castle on this side, and the gate was defended by a drawbridge which worked on a pivot and had a balance pit, like that described above at Caerphilly.

The gateway passage is in two sections, the first leading to the north

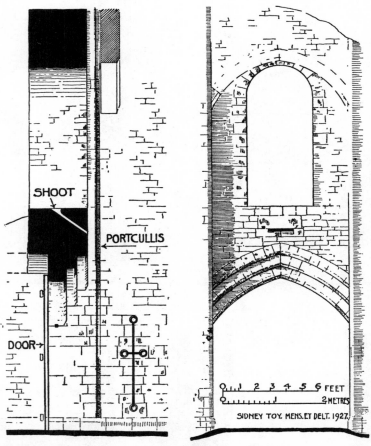

Leybourn Castle. Section and Elevation of Gateway. p. 192.

side of the central hall, a large octagonal apartment vaulted in stone, and the second from the west side of the hall to the inner bailey; thus involving a right-angled turn within the hall (pp. 195, 200). The outer section was defended by two portcullises and two doors. Those assailants who had passed these first barriers, below the machicolations, which the destroyed vaults doubtless contained, and had penetrated into the hall, found them-

GATEHOUSES AND DEFENCES OF THE CURTAIN 195

selves under attack from five meurtrières in the surrounding walls, one of which, in the south tower, faced directly towards the entrance passage. The hall passed, the inner section of the gateway, now largely destroyed but probably defended in a manner similar to the outer section, had then to be carried.

But the defensive principles of mediæval gateways reached their cul-

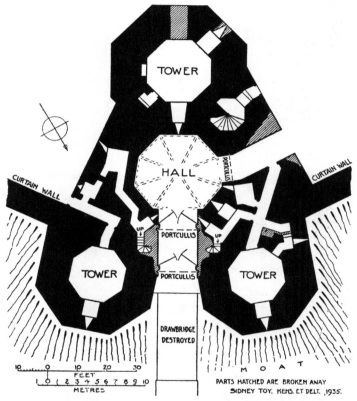

Denbigh Castle. Plan of the Gateway. p. 193.

minating point in the King's Gate at Caernarvon, built to its existing extent 1316–1320, but probably never completed. In essence the design of this gate is similar to that at Denbigh, with two passages at right angles to each other and a large hall at the junction, but the defences are more numerous and the passages longer than at Denbigh. This gate also was approached by a drawbridge working on a pivot and having a balance pit (pp. 158, 196, 200).

The first passage was defended by four portcullises and two doors and

196 CASTLES

was commanded from above by seven lines of machicolations. The outermost machicolation covers the head of the first portcullis, so that, when required, the portcullis could be drenched with water. From a point about a third of the way through, the passage gradually opens out, both in width and height, to another point about two-thirds of the way through, when it becomes parallel again. The passage was under attack from numerous

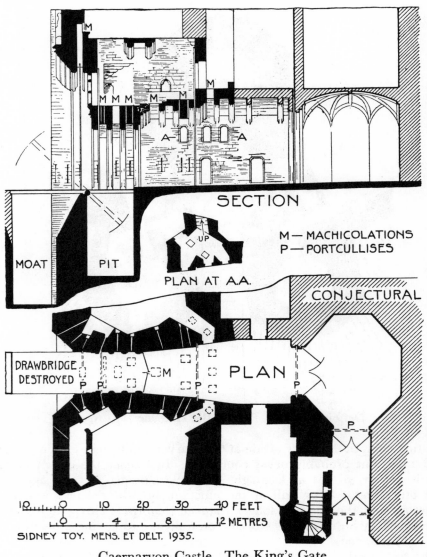

Caernarvon Castle. The King's Gate.

GATEHOUSES AND DEFENCES OF THE CURTAIN 197

meurtrières on either side, and its middle portion, from a level 12 ft. above the ground, was commanded by six doorways, three on either side, through which heavy missiles could be thrown down on assailants. The chambers behind these doorways are approached by steps down from the first floor of the gatehouse.

Only one side of what, apparently, was to be an octagonal hall, and only

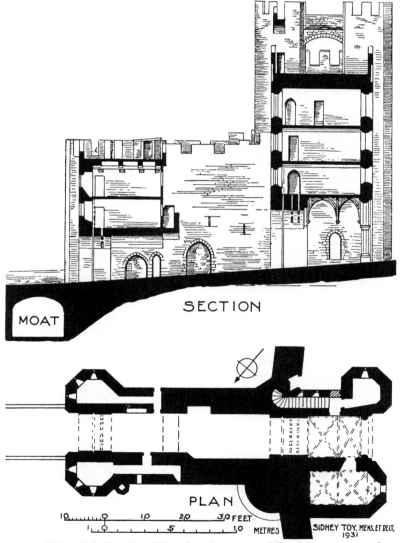

Warwick Castle. The Gatehouse and Barbican. p. 198.

one side of the second passage, leading towards the inner bailey, exist. But these fragments clearly indicate the design and show that the inner passage was to be closed by a portcullis and door at either end. So that throughout the whole gateway there were to be six portcullises and four doors.

The gatehouse at Warwick Castle, built during the latter part of the fourteenth century, is defended by a barbican which forms an extension of the gateway (pp. 197, 199, 200). The barbican consists of a forebuilding, rising two storeys above the gateway and two lofty walls which flank an open court between the forebuilding and the gatehouse. The gateway, with a rapid rise from the exterior to the interior, passes straight through from end to end and is defended by one portcullis at the entrance to the barbican and by one portcullis and one door at the entrance to the gatehouse. There are also three lines of machicolations, one on the inner side of the first portcullis and two between the second portcullis and the door; there are none before either of the portcullises. But the outer portcullis is commanded by the flanking turrets of the barbican and the inner one by the battlements round the court.

The principal defence of the passage was from the battlements of the barbican and of the gatehouse round the open court; and from a gallery in the barbican at first-floor level facing towards the court. The gallery and the upper floors and battlements of the barbican are approached from the gatehouse, and also directly from the bailey, by way of a door in the curtain and a mural passage in the south wall of the barbican.

The gatehouse rises three storeys above the gateway and has four turrets, one at each corner. The turrets are carried high above the roof and are connected by single arched bridges with parapets. There are therefore three tiers of battlements; one, the lowest, being on the roof and lateral walls of the barbican; the second on the roof of the gatehouse; and the third, made continuous by the bridges, at the summit of the turrets; and all three are in direct communication with each other by spiral stairways. There is a doorway to the wall walk of the curtain on either side of the gatehouse; but, as at Cæsar's tower, the way across from one side of the curtain to the other was over the roof and not through the gatehouse. The wall walks of the curtain on either side of the gatehouse are paved with large stone slabs.

In Scotland and the Border counties the gateways, particularly those of the smaller castles and towers, were often defended by iron gates, called yetts. Yetts were powerfully forged grills formed of vertical and horizontal bars, hung on hinges, and secured either by long iron bars drawn out from a socket in the wall or by bolts attached to the yett. There were two methods of construction. In Scotland the bars composing the gate were so forged that the vertical and horizontal pieces penetrated or formed sockets for the others

Warwick Castle. The Gatehouse and Caesar's Tower from within the Bailey.   pp. 178, 197.

Denbigh Castle. The Gatehouse. p. 193.

Caernarvon Castle. The King's Gate. p. 195.

Warwick Castle. The Gatehouse and wall walk on the curtain from the South-East. p. 198.

GATEHOUSES AND DEFENCES OF THE CURTAIN 201

in alternate series. Also the grille work was left open. In England all the vertical bars passed in front of the horizontals and the joints were riveted

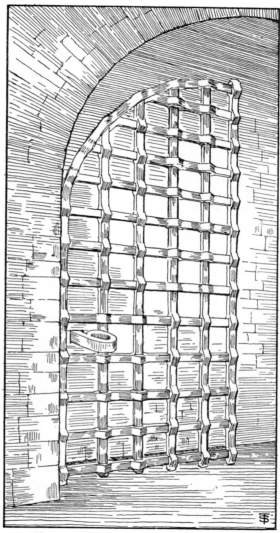

Doune Castle. Leaf of Iron Yett.

and clasped alternately. The spaces between the verticals were filled in with oak boards so that the yett was solid.

At Doune, Perthshire, dating from the latter years of the fourteenth century, the main gateway was defended by a portcullis and a yett at the entrance, and, apparently, by another yett, now missing, at the end towards the bailey. Each yett was of two leaves, hung on hinges $2\frac{3}{8}$ in. diameter; that at the entrance still remains. It is made of $1\frac{3}{4}$ in. by 1 in. bars, crossing each other in meshes $7\frac{3}{4}$ in. square, and was secured when closed by an iron bolt, $2\frac{1}{4}$ in. square. This heavy bolt was drawn out from a socket in the wall by means of a handle and stretched across behind the yett in the same manner as the timber bars behind an ordinary door. One of the leaves contains a small wicket gate.

DRAWBRIDGES

Drawbridges were operated in several different ways. Some of them, as at the Burg Ez-Zefer, Cairo, appear to have been simply drawn back upon a platform before the gate. Others as at Conway were hinged on the inner side and, by means of chains attached to the outer side, and by pulleys, were raised up until they stood vertically against the face of the gate and formed an additional barrier to the passage. This was a very common method. Drawbridges working on a pivot and having a counter-balance working in a pit have been described above.

Another method, introduced about 1300 and much employed in France and Italy in the fourteenth and fifteenth centuries, was by means of long beams mounted over the entrance arch, one on either side, at the base of tall vertical recesses (p. 203). The beams worked on a pivot at their centres and projected half outside and half inside the gateway. Chains connected their outer ends with the outer side of the bridge, and the latter, being hinged on the inner side, was raised by means of counterbalances on the inner ends of the beams, and no pit is required. When the bridge reached a vertical position it fitted into a square headed recess, and the outer ends of the beams fell back into the vertical recesses above them. When, as was frequently the case, there was a small doorway for foot passengers on one side of the main archway, the doorway was provided with a separate bridge and counterbalanced beam.

CURTAIN WALLS AND WALL TOWERS

Curtain walls and wall towers were now usually built with deep battered plinths, for greater stability and additional protection against sapping operations. Sometimes the plinths were of great thickness at the base, ascended to half, or more, the height of the wall, and were continuous round the curtain and the towers, as at Angers and Sarzanello. At Le Krak des Chevaliers the south and west walls of the inner bailey, which are more exposed than the other walls, have massive plinths so thick at the base that they project much beyond the outer faces of the wall towers, and, in rising, cover the lower parts of those towers (p. 136). They terminate at a level more than three-quarters the height of the curtain.

GATEHOUSES AND DEFENCES OF THE CURTAIN

Wall towers with prows, or spurs, rising from base to parapet, have been noted on page 100. At Visby the wall towers, dating from the thirteenth century, are square for about half their height and octagonal above that level, the corners terminating in pyramidal stops (p. 172). Often the towers were built square at the base and round or semi-octagonal above, thus having the double advantage of a widespread solid base, difficult to sap, and an upper surface with no sharp corners to hide the sappers. In earlier work the corners of the base were adjusted to the upper surfaces by deep triangular splays, as at

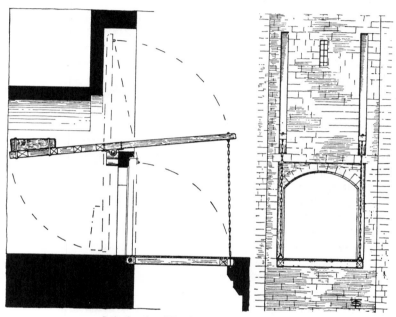

Method of Raising Drawbridge.
Section and Elevation. p. 202.

the south-west tower at Chinon, dating from the second half of the twelfth century. But at later periods the angles were finished with pyramidal spurs, as at Marten's Tower, Chepstow, built about 1240 (p. 211). Later in the century the spurs sometimes rose to within a few feet of the parapet, as in the towers of the screen wall at Caerphilly, about 1275 (p. 211), and the south-east tower at Goodrich Castle, Hereford, about 1300. During the fourteenth century wall towers along the face of the curtain were often square through their full height, as at Avignon and in the fortification of the Templars at Malta and Rhodes, though at corners and strategic points they were still more generally round or octagonal.

Watch towers, or bartizans, commanding the field through loopholes in many directions, were often built out on corbels at the angles of curtain walls or towers at the level of the wall walk.

BATTLEMENTS

The wall walks on the curtain between the towers and on the towers themselves were defended on the outside by crenellated parapets and often had low walls on the inside. The merlons, or projecting portions of the parapet, behind which the defenders were secure from attack, were normally from 6 ft. 6 in. to 9 ft. high and 5 ft. to 6 ft. wide. The embrasures were from 2 ft. 3 in. to 3 ft. wide and had a breast wall about 3 ft. high; through these apertures arrows were shot and missiles hurled at the attacking forces. From about the end of the twelfth century the merlons were often pierced in the middle by loopholes.

In some examples of the thirteenth century and later the coping of the merlons was finished with a roll, to prevent arrows which had struck the lower part of the slope from glancing over the parapet. The projecting water drips at the base of the coping also served the same purpose. At the Eagle Tower, Caernarvon, where the parapet is 9 ft. 4 in. high and there is a roll at the top of it, the projection at the base of the coping is carried down the sides of the merlons, thus protecting the embrasures from arrows glancing along the sides of the parapet (p. 167).

In times of siege the circulation of the wall walk behind a parapet with frequent embrasures must always have been attended with considerable danger, and from the thirteenth century the embrasures were often covered by wooden shutters, placed on the outside. The shutters were hung from the top, worked in sockets on either side, and opened out from the bottom. Even when opened sufficiently wide to permit attack on the enemy operating below they still formed adequate protection to the defence forces. The sockets were so made that the shutters could be lifted up and removed in time of siege, when the embrasures would be required for access to the hoarding outside. At the barbican, Alnwick Castle, Northumberland, the sockets were cut in the stonework, the trunnion fitting into a hole on one side and into a slot on the other (p. 205). When the wall walk was roofed the shutters were sometimes made in two parts, the upper part being kept slightly open for light and ventilation, and the lower part, the largest, opened, or removed, when required (p. 205). At the upper gate of the town, Conway, two embrasures of the parapet, facing each other over the entrance to the gateway, are each shielded on the outside by a large and thin slab of stone, which projects out at right angles from the face of the adjoining merlon.

HOARDS AND MACHICOLATED PARAPETS

As already shown, hoards were employed from the remotest times. They were also in general use during the Middle Ages, and some of them still exist, as at Laval, Mayenne, where the circular donjon is still surmounted by hoarding dating, probably, from the thirteenth century. In numerous other fortifications, from the twelfth century to the fifteenth century, holes for the brackets supporting the hoards are to be seen at the level of the wall walks of curtains and towers.

Hoards, or brattices as they were often called, were temporary wooden galleries, constructed on the outside of parapets in the time of siege to protect the bases of the walls and towers against sapping operations. These

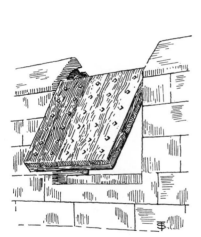

Single Shutter. p. 204.

Double Shutter. p. 204.

galleries were built upon rows of beams; each beam being about 10 in. square and long enough to stretch across the wall walk, pierce the parapet, and project as a bracket about 4 ft. beyond the outer face. Boards were laid across the brackets to form a foot-pace, a wide aperture being left for a machicolation, and the gallery was protected on the outside by a screen and covered by a pent roof.

Extensive buildings of timber on the walls being a constant source of danger, and of disaster, from burning missiles, hoards were being replaced, gradually, by stone machicolations. The first step in this direction was to build, immediately below the parapet, rows of stone corbels on which the hoarding could be built when required, as was done about 1240 in the donjon and towers at Coucy.

Towards the end of the thirteenth century machicolated parapets of stone were built at the most vulnerable points, such as the gateways, while provision was still made for hoards in other parts of the castle. At Conway Castle, 1283–87, the walls containing the east and west gateways were surmounted by machicolated parapets from end to end, while the towers and lateral walls contain beam holes for hoards. But it was not until the fourteenth century that machicolated parapets were in general use. The walls built at Carcassonne about 1285 have plain parapets with holes for hoards. In some cases, as at Caernarvon and Harlech, there does not appear to have been any provision even for hoarding, reliance being placed on the power of the wall towers to defend themselves and sweep the panels of wall between them.

At Sarzanello, about 1325, the walls and bastions all round the castle, and the walls of the outwork are defended by machicolated parapets, arranged in groups of seven or eight with short intervals between each group. And at Avignon, some forty years later, machicolated parapets were built on the wall towers of the town, and on the curtain, where they were carried along without break from tower to tower.

At the Château de Tiffauges, Vendée, there is a powerful wall tower, the Tour du Vidame, of about 1500, which has a machicolated parapet and, rising directly from the crest of the parapet, a high pitched stone covering (pp. 207, 208). Here the wall walk is really a mural gallery, roofed with flat stones, and having on one side the parapet with its machicolations and on the other a continuous stone seat with foot step, for the use of those defending the tower.

The acoustic properties of the gallery are noteworthy. Two persons seated on the bench, one at either end, having their backs to the wall and speaking in a low tone, can hear each other distinctly. The present gable end and timber roof, built for the protection of the vaulting of this invaluable tower, are modern.

In some fortifications the base of the curtain wall was defended on the outside by a vaulted passage, or casemate, which ran along from tower to tower above the inner edge of the moat and was pierced with loopholes to the field. At Gisors the foot of the curtain was originally protected by breastworks, but in the fourteenth century the breastworks were replaced by casemates, a section of which still remains on the east side of the castle (p. 62.) At the Château de Domfront, Orne, the wall opposite the town was actually a revetment, built against the face of the steep rock of the ditch, and casemates were formed in it by cutting a gallery along the face of the rock behind the revetment and piercing loopholes through the revetment itself.

SPIRAL STAIRWAYS

During the eleventh and the greater part of the twelfth century stairways in keeps, towers, and gatehouses were built upon spiral vaults, winding round a central newel. The vaults were constructed on timber centering, which was probably moved upwards as the work proceeded; and the steps, composed

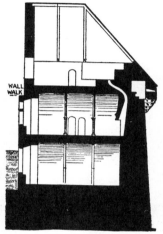

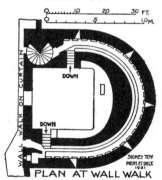

Château de Tiffauges.
Tour du Vidame. p. 206.

of flat stones or bricks, were laid upon the upper surfaces of the vaults. By this method of construction, while relatively small pieces of material could be used for the steps, the width of the stairway itself was unrestricted. But the method absorbed a considerable amount of time and the urgent demands of military building required a more expeditious process.

From about the end of the twelfth century spiral stairways were composed entirely of a series of steps, each step being cut out of one stone and sufficiently long to form a section of the newel at one end and to tail into the wall at the other. The first steps are built upon solid masonry; the others, as they rise, are supported by the newel, the edge of the step below, and the wall, and no vault is required. Later spiral vaults were again introduced, and the treads built of bricks, as at Nether Hall, Essex, about 1470, and Kirby Muxloe, Leicester, 1480–84.

By far the greatest number of these spiral stairways turn on the right as

Château de Tiffauges.
Battlements of the Tour du Vidame. p. 206.

they ascend; so that while those defending them from above have the greatest space in which to use their sword arm, assailants mounting would be at great disadvantage in this respect. But since in a conflict there must be many occasions when the positions would be reversed, there are generally some stairways turning left as they rise. In the inner bailey at Caerphilly seven stairways turn right and two left. At Conway seven turn right and one left. At Beaumaris out of ten stairways four turn left, and at Caernarvon seven turn right and four left.

GATEHOUSES AND DEFENCES OF THE CURTAIN

The number of fighting men quartered in a castle varied greatly from time to time and depended on conditions of peace or war, on the status and means of the commander and on the faithful performance of military obligations. The permanent garrisons provided for the royal castles of Caernarvon and Conway in 1284 were as follows. For Caernarvon, in addition to the constable, there were to be two serjeant horsemen, who had charge of the castle in the absence of the constable, ten serjeant crossbowmen, a smith, a carpenter, an artillery craftsman, and twenty-five footmen at arms—forty men in all.[1] For Conway, in addition to the constable and his household, there were to be thirty fencible men in all, consisting of fifteen crossbowmen, a chaplain, a smith, a carpenter, a mason, an artillery craftsman, and ten others—janitors, watchmen and other ministers of the castle.[2] In 1401–1404 the garrison provided for Caernarvon Castle, in addition to the constable, was a hundred men, consisting of twenty men-at-arms and eighty archers; Conway Castle, seventy-five men, consisting of fifteen men-at-arms and sixty archers; Harlech Castle, ten men-at-arms and thirty archers; and Beaumaris Castle, fifteen men-at-arms and one hundred and forty archers.[3]

---

[1] *Welsh Roll Chancery*, 12 Edw. I, 1284, Memb. 5; Cal. Rot. Wall. 288.
[2] Ibid. Memb. 2; Ibid. 292.
[3] *Acts of The Privy Council*, Vol II. Henry IV. pp. 64–66.

CHAPTER XVI

DEVELOPMENT OF THE TOWER-HOUSE

IN the design of castles from the latter part of the fourteenth to the end of the fifteenth century the military and domestic elements come more and more into sharp contrast. For while there is a general tendency towards the strengthening of the curtain and the outer defences, there is also an ever-increasing desire to expand the hall, the living-rooms, and the offices, and to place these domestic buildings in convenient relation to each other.

During the latter part of the fourteenth century castles built on hills, as Pierrefonds in France and Cesena in Italy, differ as much in plan as the sites they occupy vary in physical character. Those on level ground were normally rectangular, were defended by corner, and sometimes intermediate, towers, and were surrounded by moats, as the castles at Ferrara and Mantua in Italy, Vincennes in France, and Bodiam in Britain. But in all cases, whether on the hill or on the plain, great attention was paid to the approaches, which were made as difficult, dangerous, and exposed to attack from the castle as possible.

The Château de Pierrefonds, Oise, 1390–1400, stands on a land promontory and is protected on three sides by the natural escarpments of the hill. On the fourth side it is defended by a powerful outwork and, between the outwork and the castle, a wide ditch. The curtain, roughly rectangular in plan, is defended by round towers, one at each angle and one in the middle of each side; the intermediate tower on the south, defending the gateway, being much larger and stronger than the others. The towers and walls all round the castle are surmounted by machicolated parapets. There are two ways of approach, one on either side of the château; that on the east being by means of a ramp, and that on the west by a ramp and a circuitous course round the château to join the other, both then leading to a gate in the outwork. During the whole course both routes were under direct attack from the walls of the main building or the outwork. From the outwork a bridge led across the moat to the principal gate.

The Castello della Rocca at Cesena, near Forli, dating from 1380, is built on a hill at the south-west side of the city. It consists of a polygonal inner bailey standing on the top of the hill and an outer bailey which runs down its south-eastern slopes. The inner bailey is surrounded by powerful walls with towers at the angles. It encloses a large rectangular building of three storeys, and a square tower, the tower standing slightly apart from the main building

Chepstow Castle.   Marten's Tower.   p. 203.

Caerphilly Castle.   Towers of the Screen Wall.   p. 203.

Tantallon Castle from the South-East.

and both clear of the walls. The gatehouse, projecting within the curtain, is on the north side of the bailey and there is a postern on the south side; both opening on to the outer bailey.

The gatehouse to the inner bailey is strongly defended. It is flanked on one side by a large round tower, built slightly outside the bailey but joined to it by a short wall, and on the other side by a smaller tower and the curtain wall of the outer bailey, which runs down the hill parallel with the line of approach. In front of the gatehouse there is a small barbican, with its entrance at right angles to the main gateway, and the approach to the barbican is intercepted by a series of cross walls, built between the round tower and the outer curtain and forming a maze-like passage with gateways at the ends of the cross walls. Here an enemy, even after he had climbed the hill in the teeth of attack from the battlements of the outer curtain and before he had reached the barbican, must negotiate this serpentine passage, exposing each flank in turn, and open to enfilade throughout the whole course.

Tantallon Castle, East Lothian, built probably during the third quarter of the fourteenth century, stands upon a bold promontory, jutting into the sea, and is protected on three sides by precipitous cliffs. On the fourth, the land side, it was defended by one and probably two baileys, and, though the outer ditch may be the result of siege operations, by three lines of ditches, all drawn across the promontory from the cliff on the north to the cliff or a deep ravine on the south. The approach to the middle bailey was further defended by a ravelin which projected into the outer bailey (pp. 212, 214, 183).

The inner bailey is at the point of the promontory, cut off from the rest of the works by a massive screen wall, built across from cliff to cliff; the wall having a gatehouse in the middle and a round tower at either end. Along the edge of the cliff on the other sides of the bailey domestic and military quarters, of less substantial character than the screen wall, were built. These latter, except for a range on the north side, containing a large hall, a kitchen, and other offices, have been destroyed, or have fallen down. The outer portions of the towers flanking the great wall have also been demolished, and the gatehouse, which was defended by a barbican, has been subjected to alterations and additions; but the wall itself, which is over 12 ft. thick, remains intact to the height of the wall walk and retains portions of the parapet.

The strength of Tantallon lay largely in its inaccessibility; being defended on the land side by its powerful screen wall and its outworks, and on the other sides by precipitous cliffs, rising sheer out of the sea.

In mountainous countries the castle often consisted of a single tower, perched on the top of a hill, and surrounded by defensive works so designed as to make the tower as inaccessible and its approaches as dangerous as possible, as at Troyenstein in the Austrian Tyrol dating probably from the latter part of the fourteenth century (p. 215).

214   CASTLES

Here the tower is surrounded at the base by a spiral path, enclosed by walls on either side and defended by battlements which stand high above the path. So that an enemy who had rushed the entrance gate, in face of fire from three tiers of battlements, would be exposed to attack on both sides, during the whole circuit of the path, from the gateway to the upper platform. Arrived at this point, the entrance doorway to the tower stands so high above the platform that it was only to be reached by means of a rope and basket let down from the staging before the doorway. Finally, as a last resource, if all those obstacles had been overcome and the enemy had got possession of the tower its occupants could escape by way of an underground passage at the base of the tower.

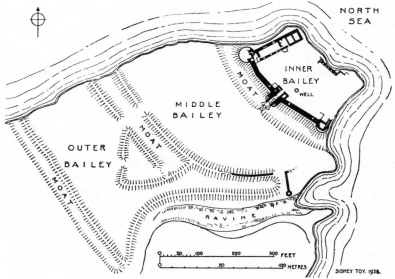

Plan of Tantallon Castle. p. 213.

The Castello d'Este at Ferrara, dating from 1385, and the Castello di Corte at Mantua, 1395–1406, are both built on level ground and are surrounded by wide moats. They are rectangular castles with a square tower at each corner and their living quarters and offices are built all round an internal courtyard. Since they each stand within the walls of an important city their approach works are restricted, but even here the approach to each gateway, as originally disposed, was by way of two drawbridges and an outwork standing in the moat between the bridges. Both castles have the widely-spread high plinths common in Italy, and the walls and towers, where not altered at a later period, are defended all round by machicolated parapets. At Mantua three of the corner towers preserve their original crenellations with

V-shaped notches in the merlons, a feature, apparently more decorative than useful, often employed in Italy in the fourteenth century.

One of the most powerful fortifications of this period was the Bastille at Paris, built 1370–1383 on the site of the Porte, or Bastide Saint-Antoine and destroyed in 1789. It was an oblong structure of great height, strengthened

Troyenstein Castle. After Otto Piper. p. 213.

by eight powerful towers, one at each corner and two in the middle of each of the long sides, and surrounded by a wide moat. There was, originally, a gateway on each side, defended by two of the wall towers and approached from across the moat by one or two drawbridges. The walls and towers were of the same height and the machicolated parapets were continuous. Here the flat roofs of the towers formed spacious fighting platforms, commanding the city on the west as well as the field to the east, and since the wall walk was continuous all round the castle, troops and engines could be rushed from point to point with the greatest facility.

Bodiam Castle, begun in 1386 and finished probably about 1390, stands on level ground on the banks of the river Rother. It has a rectangular plan with a drum tower at each corner, a gatehouse flanked by square towers in the middle of the north side, and a square tower in the middle of each of the other sides; there is a postern in the intermediate tower on the south (pp. 217, 220).

The castle stands in a large rectangular moat and is defended on the north by two advanced works, a barbican and an octagonal outwork, both standing isolated from each other in the same moat as the castle. Communication between them was by means of drawbridges. Approach to the castle was by way of a timber bridge which spanned the moat between an abutment on the west bank and the octagonal outwork. So placed, the bridge was at right angles to the gateway and stood exposed to flank fire from the walls of the castle; it was entered from the bank by a drawbridge. To reach the barbican from the octagon it was necessary to turn to the right and cross a second drawbridge, and from the barbican a third drawbridge led to the main gateway of the castle. From the postern another long timber bridge, with, probably, a drawbridge at either end, led across the moat to the south bank. Both bridges have been destroyed, and the castle is now approached by a causeway, built probably in the sixteenth century, which juts out towards the octagon from the north bank of the moat.

The gateway is defended on the outside by loopholes with oillets for firearms and by machicolations at the parapet of the gatehouse. Within the passage there were three portcullises and three doors, the innermost portcullis and door being closed against the bailey. One of the mediæval portcullises, which is made of oak, plated and shod with iron, still remains. In the vaulting, both of the main gateway and the postern, the stone bosses are pierced with round holes about 6 in. diameter, the underfaces of the bosses being about 2 ft. 6 in. below the floor line of the rooms over the gateways. These holes, dispersed over the vault and camouflaged by the bosses, were for attack on enemies entering the gateway. Only the towers of the gateway and postern have machicolated parapets; elsewhere the parapets have plain crenellations.

Within the castle the domestic quarters are ranged round the walls between the curtain and a rectangular courtyard. On the south are the great hall, the buttery, pantry, and kitchen; on the east, opening from the great hall, the private and guest halls and the chapel; on the west, adjoining the kitchen, the servants rooms and other domestic offices; and on the north, the military quarters. The postern opened from the "screens" in the great hall.

DEVELOPMENT OF THE TOWER-HOUSE

The freedom now exercised in the disposition of the domestic buildings within the castle became a source of danger. At Bodiam none of these buildings could be held against an enemy who had penetrated into the

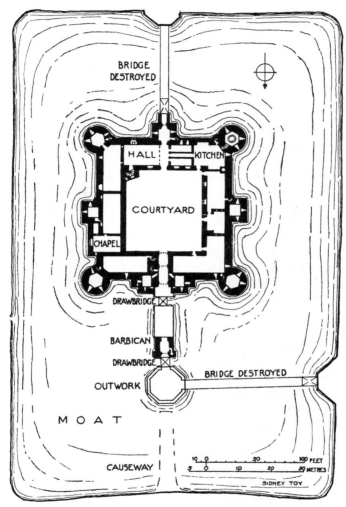

Fig. 136. Plan of Bodiam Castle.

courtyard, or against mercenaries, often employed by leaders at this period and lodged within their castles, who had become disaffected or treacherous. It soon became obvious, probably from painful experience, that greater security for the principal dwelling rooms was essential.

Already, at Vincennes, begun in the first half of the fourteenth century, the donjon containing the royal quarters was isolated and strongly fortified; and at Dudley the residential keep built at this period was an independent structure, defended as well against the bailey as the field.

At Pierrefonds, 1390–1400, the quarters of the retainers and garrison are ranged round the courtyard inside the curtain, and, as became the status of the powerful prince, the Duke of Valois, who built them, they are designed on a palatial scale. But the particular residence of the Duke is a tall rectangular structure near the gateway, many storeys in height, separated from the other buildings, and capable of independent defence. The tower-house of Warkworth Castle, Northumberland, built about 1390 on a mount on the opposite side of the bailey from the gatehouse, is also an independent fortified residence, standing apart from the other buildings, and commanding both the field on the north and the bailey on the south.

Doune Castle, Perthshire, about 1395, is designed on principles similar to those of the inner bailey at the Château de Pierrefonds, though on a more modest scale. And the analogy is all the more striking in that the entrance gateway in both castles is defended by a half-round tower incorporated in the tower-house, though at Pierrefonds the gateway passes beside the tower-house while at Doune it passes through it. The tower-house at Doune is at the north-east corner of the bailey, which is roughly rectangular in plan, the great hall extending west from it, and the kitchen and offices opening from the south-west of the great hall. Other buildings, since destroyed, were ranged round the east and south walls of the bailey and there is a well in the courtyard (pp. 219, 201).

The ground-floor rooms of the tower open off from the gateway, and were for the use of the guard; they also contain a prison and a well chamber. The entrance to the upper rooms was by way of a flight of steps up from the courtyard to a doorway on the first floor (p. 219). Another flight of steps, to the west, led to the great hall and offices. At present a way has been broken through between the first floor of the tower, called the Baron's Hall, and the great hall; originally there was no communication between these two rooms, but there was a means of approach to the great hall from the hall on the second floor of the tower by a spiral stairway. The door of entry from the tower to this stairway, however, was closed against the great hall. So that while normally the lord of the castle probably used the great hall for dining, since his service otherwise would have been restricted, in case of necessity the door on the stairway could be closed and barred on the inside and the tower isolated from the rest of the castle.

In many castles built during the fifteenth century the domestic quarters are contained in large rectangular towers, which were either incorporated in the curtain wall, or stood isolated within the bailey. Generally these towers could be defended independently, and some of them, particularly those in

Doune Castle from the North-East.   p. 218.

Doune Castle. The Tower House from within the Bailey.   p. 218.

Elphinstone Tower from the South.   p. 224.

Bodiam Castle from the North.   p. 216.

Bodiam Castle. The South and West Ranges of the Courtyard.   p. 216.

DEVELOPMENT OF THE TOWER-HOUSE 221

Spain and in Scotland, are of great strength. Many of them were well-lighted and conveniently planned dwelling houses. Borthwick Castle, Midlothian; Tattershall Castle, Lincoln; and the Castillo de Fuensaldaño, Valladolid, are examples.

At Borthwick the residential quarters are all contained in a great stone tower which stands isolated within the bailey. (See also p. 222). The castle, built 1430–1440, stands on a promontory at the junction of two streams and is wedge shaped on plan; the long sides, on the north and south, meeting at the point of the promontory. The gateway, now rebuilt, is at the

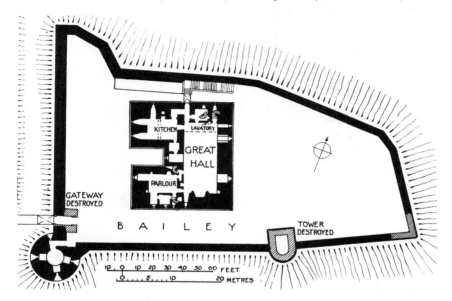

Plan of Borthwick Castle.

south end of the west wall, which forms the base of the wedge, and was defended by a drum-shaped corner tower.

The great tower within the bailey, while being designed as a convenient residence is also a very powerfully-built structure, vaulted at various stages and at the summit. It consists of a main rectangular body and two wings, both wings projecting out from the west side and rising to the full height of the tower. There are two entrances. One of them is at first-floor level, and is approached by a bridge thrown across from the wall walk of the curtain; the wall walk being reached by a flight of steps up from the courtyard. The other entrance is immediately below the first and leads from the courtyard down to the basement, from which level two spiral stairways rise to the first floor.

The first floor contains the great hall, the parlour, and the kitchen; the parlour opening from the dais at one end of the hall and the kitchen from the "screens" at the other end. Spiral stairways from the parlour and from one end of the screens rise to the upper floors and the battlements. In the kitchen wing there is a special service stairway. On one side of the screens is a beautiful lavatory, with carved canopy and wall shaft below the basin, similar in design to the piscinæ in some churches. Such, in brief, are the

Borthwick Castle. The Tower from the North-West.

domestic arrangements of this building. The tower is surmounted by a machicolated parapet which ran all round the walls and is capable of offering strong independent resistance.

The tower at Tattershall was built in 1433-1443 on the west side of the inner bailey of a thirteenth century castle (p. 223). The old curtain wall was taken down at this point and the tower, constructed of brick with stone dressings and base, built projecting out into the moat with one of its long sides in line with the curtain.

Of the fortifications of the inner bailey eastward of the tower little original stonework remains save the foundations of two round towers, one on

DEVELOPMENT OF THE TOWER-HOUSE

either side of the great tower, and other foundations and fragments of masonry in line with the curtain. Foundations of kitchens, added in the fifteenth century, are also to be seen to the south of the tower.

The tower consists of a vaulted basement, a ground floor and three upper floors, and has at each corner an octagonal turret, which rises from the base to a level high above the walls. Each storey above the basement, which was

Tattershall Castle. The Tower from the South-East.

used for stores and contains a well, has a large central hall with small chambers opening from it, the hall being lighted by relatively large windows. There are three entrance doorways, all opening from a passage or forebuilding, now destroyed, which stood between the tower and an earlier hall. One doorway leads by a flight of steps down to the basement, the second straight in to the ground floor and the third to a turret stairway leading to the upper floors and battlements. The purely residential quarters, with communication to the kitchens were on these upper floors.

The turrets are surmounted by simple crenellated parapets but the walls between the turrets have two tiers of battlements, the lower tier having a machicolated parapet. The embrasures of the lower tier were originally covered by wood shutters, hung from the top.

Though the tower at Tattershall has strong defences at the summit, personal comfort rather than military strength is the dominating factor in its design. Its windows are large, all the floors above the basement are, and always were, of timber, and the defences at the entrance doorways must have been remarkably weak.

The Castillo de Fuensaldaña in Spain consists of a rectangular bailey, enclosed by a curtain with wall towers, and a great rectangular tower, which stands at one end of the bailey and projects partly within and partly outside the curtain. The whole was built in the fifteenth century. The gateway to the bailey passes between the great tower and a drum tower at one angle of the curtain wall and is commanded by both of them. Here also the great tower could be defended independently of the rest of the castle. It is approached by way of a square newel stairway and a bridge thrown across from the top landing to a doorway in the tower, the building containing the stairway standing isolated within the bailey.

Many towers with no other defences than a barmkyn, or walled courtyard, and differing little in character from those of the previous century, were built in Scotland at this period. In these towers there is now a tendency to increase the number of mural chambers, to divide the upper rooms by partitions, and to bring the kitchen into more convenient relationship with the great hall. Elphinstone, East Lothian, and Newark, Selkirkshire, both rectangular buildings without wings, and both dating from the fifteenth century, are examples.

Elphinstone tower is of four storeys, the second and third being vaulted and the fourth covered by a timber roof. Both the third and fourth storeys are lofty, well-lighted halls with mural chambers at the corners and sides. A portion at one end of the third storey, with a wide fireplace, is partitioned off from the great hall to form a kitchen, and the fourth storey was also divided into two rooms by a partition. The entrance doorway is placed at a level between the first and second storeys, only a few steps above the ground. But the doorway gives direct access to the lower storeys only. To reach the great hall and upper floors it is necessary to ascend a flight of steps, rising up through the wall from the entrance passage and commanded from above by an opening in the wall of the great hall. From the great hall spiral stairways in the corners of the tower rise to the upper rooms and one of them to the battlements (p. 219).

Newark Castle, standing on the crest of a steep bank of the Yarrow water, consists of a large tower, built in the early years of the fifteenth century, and a barmkyn, the wall of which, now in ruins, probably dates from the latter part of the sixteenth century (p. 227). The tower measures 65 ft. by 40 ft.

externally and is six storeys in height, the uppermost storey being formed partly in the gabled roof. The second storey only is vaulted. Here the original entrance was at the third storey and gave direct access to the great hall. The present outer doorway at ground-floor level was inserted at a later date. The entrance was defended by a guard-room, formed in the wall and opening out of the entrance passage. A spiral stairway at one corner of the tower led from this level down to the first storey and a stairway in the opposite corner down to the second storey; both stairways rising to the battlements and each terminating in a gabled turret or cap-house. The kitchen, with an enormous fireplace, is formed at one end of the great hall, and was probably cut off from the hall by a partition, now destroyed. The upper floors also appear to have been divided by partitions, each into two rooms.

In the Border counties, both in England and Scotland, strongholds, called pele-towers were built in great numbers during the fifteenth and sixteenth centuries. The constant raids for cattle and movable goods among the inhabitants of those regions made the provision of places of refuge for person and property a dire necessity. On warning being given, the people fled to the nearest pele, made their cattle secure within the barmkyn and themselves within the tower. The origin of the word pele or peel is somewhat obscure. As far as is known, the word, said to be derived from the Latin *palus* (a stake), does not occur in documents before the sixteenth century, and then usually in reference to stone structures. These buildings, varying only in size and detail, consist of a tower and barmkyn and differ in no essential point from the Border towers already described. Smailholm tower, Roxburgh, built in the sixteenth century is a typical example.

In the domestic development of the tower-house in Scotland during the sixteenth century there were wings at the corners, or sides, of a rectangular main structure, producing Z, T, and E-shaped plans.

DOMESTIC BUILDINGS

The arrangement of domestic buildings as seen at Bodiam followed a plan, which, with variations, was common in England throughout the whole mediæval period. Generally the hall was a lofty one-storey building, having large windows to the courtyard and an open timber roof. Fireplaces were placed either in the side walls, as in the Great Hall at Kenilworth, or on a hearth near the centre of the hall. In the latter case the hearth was provided with a reredos, and probably a hood, and smoke escaped through a louvre in the roof. At one end of the hall there was a raised platform, or dais, on which stood the high table. At the other end a passage, called the "screens" was formed between the end wall and a partition, either of stone or wood, which stretched across the hall from side to side and supported a minstrels gallery over the passage. The main doorway to the hall, often protected by a porch, was at one end of the screens and a doorway or postern at the other. There were generally two doorways in the partition opening to the hall, and often three doorways in the end wall, as seen at Bodiam, one opening to the pantry, another to the buttery, and the middle one to a passage leading to the kitchens (p. 217).

The buildings on either end of the hall were of two storeys; and the solar, the retiring room of the lord, his family, and his guests, was usually on the dais side of the hall and was often approached by a stairway leading out of the hall. The chapel was generally placed near the hall or the private rooms. Such was the English arrangement in essence, but even in England there were as many variations of this plan as there were castles. Generally in all European countries the kitchen was placed in a convenient position in relation to the great hall; or, where the domestic buildings were extensive, the kitchen stood near the dining hall.

The hall was often constructed on a magnificent scale, with large traceried windows and wide fireplaces, as at Pierrefonds and the Great Hall at Kenilworth, built by John of Gaunt in 1392 (pp. 75, 228). Chapels were sometimes separate buildings of considerable size and elegance, as that built in the fifteenth century at Ashby-de-la-Zouche, Leicestershire (p. 228).

PRISONS

Though the name prison is often misapplied to ill-lighted rooms which were actually stores, or even latrines, there was usually one and in large castles two or more prisons. Prison cells of Launceston and Roumeli Hissar have already been noted (pp. 59, 83-4). The basement of the south-east tower of Skenfrith Castle was a prison, entered through a trap-door in the room above and having no other outlet than a ventilation shaft, which passed steeply up through the wall to a small opening on its outside face. There are similar cells in the basement of two of the towers at Conway Castle, though here the openings are larger, pass more directly through the wall, and admitted a dim light as well as air.

The basement of Cæsar's Tower, Warwick, was a prison (p. 178). It has no communication with the upper floors of the tower but is approached by a flight of steps down from the courtyard to the prison door. From this point another flight of steps led up to a gallery between a small window and an opening, originally closed by a grill, to the prison, so that the warder could overlook the prisoners from the gallery without himself entering the chamber. The prison is paved partly by stone slabs and partly by bricks. It is provided with a latrine and is ventilated by an air shelf to the courtyard.

At the Château de Pierrefonds the two lower stages in four of the towers are prisons, all circular cells, about 13 ft. diameter and covered with stone vaults. In each case the upper cell is reached by a spiral stairway, leading down from upper floors, and a passage with two doors between the stairway and the chamber. It is lighted by two loopholes, placed high in the wall well

Newark Castle, Selkirkshire, from the East.   p. 224.

Kenilworth Castle. Interior of the Great Hall. pp. 225-26.

Ashby-de-la-Zouche Castle. The Chapel from the North. p. 226.

out of reach, and has a latrine. A circular hole with a stone cover in the centre of the floor forms the only entrance to the lower chamber. This latter, a bottle-shaped cell, is provided with a latrine but has neither light nor ventilation, and when the stone cover was replaced over the hole in its roof was little other than a living tomb. About the time of the restorations carried out under Viollet-le-Duc in 1862, the skeleton of a woman, crouched in the latrine recess, was found in one of these lower chambers.

At Dalhousie Castle, Midlothian, the prison is approached by a mural stairway down from the first floor of the fifteenth-century tower-keep, and is entered by a small doorway 1 ft. $10\frac{1}{2}$ in. wide by 3 ft. $4\frac{1}{2}$ in. high. It is provided with a latrine, having a cesspit below the timber floor of the prison, and is ventilated by an air shaft, but has no window. At Crichton Castle in the same county the entrance doorway to the prison is only 1 ft. 11 in. wide by 2 ft. 5 in high and the sill of the doorway is 6 ft. above the floor of the cell. This prison also is unlighted, but it is ventilated by an air shaft. The cell at Warwick measures 19 ft. 3 in. by 13 ft. 4 in.; that at Dalhousie 10 ft. 10 in. by 10 ft. 3 in.; and that at Crichton 7 ft. 2 in. by 6 ft. 8 in. All three are covered by stone vaults.

CHAPTER XVII

SIXTEENTH-CENTURY FORTS

WHILE the development of the living quarters on the lines of convenience was proceeding the effective use of firearms was coming more and more into prominence. Great strides in the design of artillery were made during the fifteenth century. At the siege of Constantinople in 1453 the Turks used a gun which cast shot 600 lbs. in weight. In 1494, when the troops of Charles VIII marched through Italy, the French artillery reduced fortress after fortress with astonishing rapidity. But although provision was made for firearms in castles as early as the fourteenth century it was not until much later that buildings designed especially for the use of guns were constructed. From the early years of the sixteenth century, however, there is a complete disseverance of military and domestic work; the hall and its appendages develop into the palace, or mansion house, and the military works into the fort.

The first adjustment of the walls for the use of firearms was made in the design of the loopholes, as seen at the main gateway at Bodiam castle and at the West Gate of the City of Canterbury (p. 231). The base of the hole is circular, for the mouth of the piece, while the upper part remains a vertical slot for sighting. Sometimes during the fifteenth century two circular holes were made, one at the base and the other at the head of a short slot, the lower for the piece and the upper for sighting, as at the Château de Falaise, Calvados. At Kirby Muxloe, Leicestershire, 1480–1484, the hole and slot are separated.

Later, loopholes assumed various forms. Frequently the hole was widely splayed towards the outside in order to give the gun the greatest possible lateral sweep, as in the fortifications of Périgueux, Avallon, and as inserted in the gatehouse at Caerlaverock Castle (p. 168). In situations high in the face of the wall the sills of the loopholes were sloped rapidly downwards in order to repel hostile operations near the walls, as in the Carlsthor gate at Munich, and the tower, built in 1550, at Salignano near Ortranto in Italy. In the early part of the sixteenth century the guns, now of considerable weight, were being mounted on the curtain walls and flat roof of strong towers, behind suitable embrasures; or in forts designed especially for them.

Among the earliest and most perfect examples of towers designed for defence by artillery is the Burg Muqattam in the citadel at Cairo, built probably in the early years of the sixteenth century (p. 232). This powerful tower is four storeys in height, its walls are 22 ft. thick and it is covered at each stage by a strong dome vault. It is entered from the wall walk of the curtain by a passage leading straight to the central hall of the third storey; and the uppermost floor and battlements are gained by mural stairways rising from the entrance passage.

In all stages above the basement mural chambers for the guns open off from all sides of the central hall. Each chamber is pierced on the outer wall by two rectangular openings, one above the other. The lower opening with widely splayed jambs, was for the gun and is sufficiently high to permit of sighting above the piece. The upper opening was for the admission of light and the escape of smoke. Ample space is provided in the chambers for ammunition and there is a latrine on each floor. The Burg Al-Wustany, now destroyed above the level of the curtain, is of the same date and character as the Burg Muqattam.

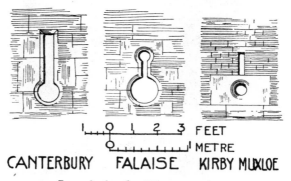

CANTERBURY   FALAISE   KIRBY MUXLOE

Loopholes for Firearms. p. 230.

Political events in England led to further developments of the fort. During the reign of Henry VIII many old castles which had been allowed to fall into disrepair were reconditioned and put into defensive order. The use of gunpowder had diminished their value but had not sufficiently developed to render them obsolete. Even the artillery of a century later when brought by the Parliamentary army against Corfe, a castle of particularly strong masonry, was powerless to reduce it, though the guns battered away at its walls for nine months.

But something more modern was necessary to meet the impending trouble. Henry had broken with the Pope and was at variance with the Emperor; he might expect an invasion at any moment. He therefore decided to build at various points along the south coast of England a line of forts, designed for artillery and equipped with the latest type of guns. Well-preserved examples of these forts, built about 1540, exist at Walmer and Deal in Kent, Camber in Sussex, and St. Mawes and Pendennis in Cornwall.

232   CASTLES

The principle governing the design of the new forts, or castles, was that the whole building should be concentrated in one compact block which could be defended all round by artillery; the guns being mounted on platforms rising

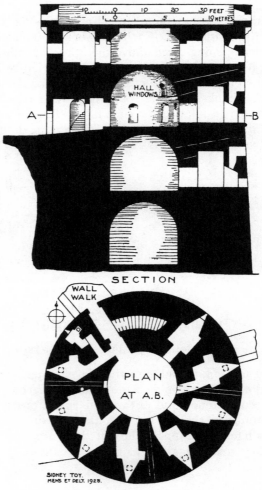

Cairo. The Citadel.
Burg Muqattam. p. 230.

in tiers one behind the other. This principle was followed in all the castles, though there are differences of plan and detail.

Walmer Castle consists of a quatrefoil-shaped curtain, or chemise, one storey in height, and a round central tower, two storeys in height, the space between the chemise and the tower being enclosed and covered with a flat roof forming a gun emplacement. A circular pier in the centre of the tower contained a spiral stairway by which the emplacement and the upper storey of the tower were approached. The whole building is surrounded by a wide and deep moat and was entered by way of a drawbridge over the moat and a doorway in one of the lobes of the chemise.

From the vaulted passage between the chemise and the tower a stairway leads down to a casemate or mural gallery, which runs all round the walls of

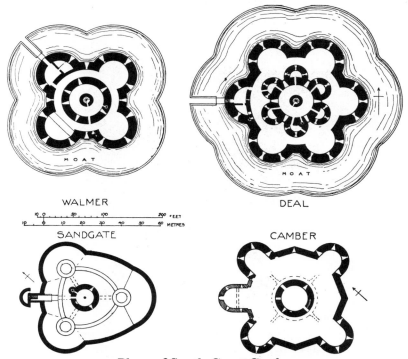

Plans of South Coast Castles.

the chemise. The gallery is pierced on the outside by numerous loopholes, fifty-six in all, from which the surface of the moat could be raked by gunfire at every point. On one side of the gallery there are large recesses for the accommodation of artillery-men and on the other smaller recesses for their ammunition. Ventilation and outlet for smoke is secured by circular holes in the vault of the gallery with shafts to loopholes on the outer face of the wall. The larger guns were mounted on the emplacement above the chemise and commanded the neighbourhood of the castle in all directions.

Deal castle is much larger than Walmer, and has two tiers of gun emplacements. It consists of a central tower and a sexfoil chemise; the tower rising to the height of three storeys and having attached to its outer face six lunettes, or lobes, which rise through two storeys only. The chemise is one storey in height and the space between it and the tower is covered by a flat roof forming the first gun emplacement. The upper floors, and battlements are reached by a spiral stairway in the centre of the tower, and, as at Walmer, the castle is surrounded by a moat and approached by a drawbridge.

The lobes attached to the tower had flat roofs for the second tier of gun emplacements; and these lobes being arranged on plan alternatively with

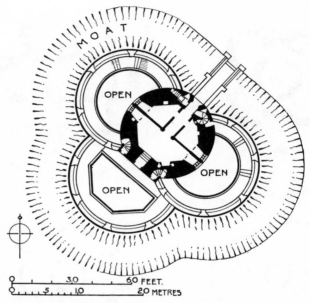

Plan of St. Mawes Castle. Based largely on plan published by H.M. Stationery Office, by permission of the Controller.

those of the chemise, the guns mounted on the two tiers were together able to direct a concentrated fire from a large number of pieces on any point around the castle within their range. Both Walmer and Deal castles have been altered and added to but the original walls remain substantially intact.

Sandgate Castle, a large portion of which has been pulled down, consisted of a central tower and two lines of triangular-shaped chemise walls, the inner line having a tower at each angle and the outer having rounded angles. It was entered by way of a D-shaped gate-tower and a stairway between the tower and the outer wall. Here the lower gun emplacements appear to have been at the rounded angles between the towers of the inner chemise and the outer wall. The upper emplacements have been destroyed (p. 233).

Camber Castle consists of a central round tower and a multangular-shaped chemise, the chemise having a large lobe projecting from each of the cardinal points. A fifth rounded projection, containing an entrance porch and having much thinner walls, was added at a later period (p. 233).

At St. Mawes there are three lobes with gun emplacements arranged in trefoil form round a central tower (p. 234); and at Pendennis a wide emplacement passes concentrically round a circular tower, except at the

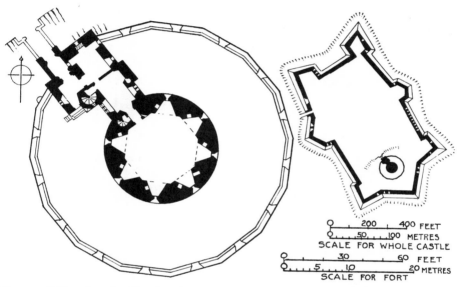

Plans of Pendennis Castle. Based largely on plans published by H.M. Stationery Office, by permission of the Controller.

point where the emplacement is intercepted by the gateway. In the latter part of the sixteenth century the fort at Pendennis was enclosed within an extensive ward, defended by a curtain wall and ditch.

One of the most powerful forts of the sixteenth century is the château Munoth at Schaffhausen, Switzerland; built at a strategic point on the right bank of the Rhine, where the river takes an abrupt bend southward. The fort is built in the form of a large circular tower, having very thick walls and strongly vaulted throughout. It consists of a basement and ground floor and guns were mounted in both of them; as well as on the flat roof. Swift and easy communication between the floors and the roof was assured by the provision of three stairways and a spiral ramp; the ramp being of such width and easy gradient as to permit of the moving of gun carriages through it. Two walls radiate from the tower down to the Rhine, enclosing a triangular barbican between the fort and the river.

Artillery now being strong enough to destroy masonry from a considerable distance it was obvious that walls had to be lowered and the line of defence brought nearer to ground level. Lofty walls and towers gave place first to lower structures, then to works half above and half beneath the ground, and, after the temporary return to early forms in the "Martello" and similar towers, finally to fortifications entirely below ground level. But the pursuit of the further development of forts, with their complicated systems of casemates and outworks is beyond the scope of this work and belongs properly to a special field of study.

INDEX

Descriptions and principal references are placed first.

ABYDUS, 17
Acre, Siege of 147
Adulterine Castles, 57
Aeneas Tacticus, 17
Aïn Tounga, 46
Africa, North, 46
Agricola's Forts, 35–6
Aigues Mortes, Tour de Constance, 128, 94
Alexios Comnenus, 144
Alfred the Great, 51
Algeria, 46
Alnwick Castle, 204
Ambracia, 27, 28
Anadoli Hissar, 83
Anadoli Kavak, 86
Anastasius, 44, 45
Angers, Château de, 135
Anglo-Saxon MSS., 51
Antioch, Siege of, 144
Antoninus Pius, Wall of, 38
Aosta, 30
Araberg Castle, 125
Archimedes, 16, 17, 18
Arques, Château de, 73
Arundel Castle, 61, 52
Ashby-de-la-Zouche, 226
Ashur, 1
Assyria, 9
Atchana, 1
Athens, Dipylon, 11
Attila, 48
Aurelian, 33
Autun, 34
Avallon, 230
Avignon, 171, 206

BABYLON, 1, 11
Babylon of Egypt, 43
Bagai, 47
Bailey, 52, 100, 116, 135, 149
Ballistæ, 143, 18
Bamborough Castle, 66, 102
Barbican, 10, 48, 103, 106, 153–55, 157, 191
Barca, 14
Barmkyn, 179, 224–5
Bartizan, 204
Bastide, 215
Bastille, 215
Bastion, 21, 164
Battering Rams, 13, 14, 142, 144, 147

Battlements, 204–6, 9, 14, 21, 24, 33, 45, 46, 110
Bayeux Tapestry, 52, 54
Beaks, 139
Beaumaris Castle, 161
Bede, The Venerable, 38
Beeston Castle, 136
Beffrois, or Siege Towers, 142, 14, 18, 27, 145
Belisarius, 46, 49
Berm, 36, 38
Blackberry Castle, 26
Bodiam Castle, 216–17, 225
Boemond, 145
Bonding Timbers, 22, 25, 62, 121–22, 185
Borcovicium, 36, 37
Border Towers, 179, 225
Borthwick Castle, 221
Bosporus, 83
Bothwell Castle, 130
Boulogne, Château de, 139
Bows and Arrows, 141, 13
,, Longbows, 141
,, Crossbows, 141–42
Bramber Castle, 52
Brattices or Hoards, 205, 14, 5
Bridges, 172
Brinklow Castle, 53
Brionne, Château de, 53
Brochs, 25
Burgh, 39
Burhs, 51
Bury Castle, 26
Buttery, 226, 216

CAERLAVEROCK CASTLE, 163
Caerleon-on-Usk, 35
Caernarvon Castle, 157, 195
,, Town Walls, 171
Caerphilly Castle, 153, 191–2, 203
Cæsar, Julius, 25, 27, 141
Cahors, Pont Valentré, 172
Cairo, City Gates, 104
,, The Citadel, 89
,, Burg Ez-Zefer, 106
,, Burg Muqattam, 230
Camber Castle, 235
Canterbury Castle, 72, 113
,, City, Loopholes in West Gate, 230
Cap-house, 179, 225
Capua, 16
Carcassonne, 139, 189, 206

INDEX

Carchemish, 6
Carlow Castle, 126
Casemates, 206, 233, 236
Castle Heddingham, 52, 115
Castle Rising, 76, 102
Castra Prætoria, 30
Cat, 142
Catania, 139
Catapults, 143, 18
Cataracta, 18
Cesena, Castello della Rocca, 210
Chains, 14, 142
Chapels, 102, 70, 72, 79, 226
Charlemagne, 50
Charles VIII, King of France, 230
Château Gaillard, 116, 148
Châtillon-sur-Indre, 92
Chemise, 91, 92, 232, 233
Chepstow Castle, 101, 112, 203
Chester, 35, 51
Chinon, Château de, 122, 203
Chipchase Castle, 180
Chippingham, Peace of, 51
Circular Castles, 165
Closeburn Castle, 180
Clun Castle, 66
Colchester Castle, 71, 60, 113, 114
  ,, City, 38, 51
Colmieu, Jean de, 52
Conisborough Castle, 97, 114, 115
Constantine the Great, 30
Constantinople, 41, 83, 230
Conway Castle, 157, 163, 202, 209, 226
  ,, Town walls, 171
Corfe Castle, 74, 189, 54, 57, 114, 116, 171, 231
Coucy, Château de, 128, 205–6, 94, 134
Cow Castle, 26
Craigmillar Castle, 180
Cranes, 19
Criccieth Castle, 161
Crenellations, 204–6, 9
Crichton Castle, Prison, 229
Crossbows, 141–42
Crusades, 90, 116, 144–48
Ctesias, 1
Ctesibius, 27
Curtain Walls, 12, 21, 202, 100, 116–17, 38, 41, 45, 58, 85, 159
Cyclopian masonry, 3

DALHOUSIE CASTLE, Prison, 229
Danube, Forts, 48
Dapour, Hittite City, 5
Dara, 45
Dardanelles, 49
Deal Castle, 233
Demetrius Poliorcetes, 13
Denbigh Castle, Gatehouse, 193
Devizes Castle, 101
Diades, 14
Ditches, or Moats, 1, 16, 22, 25, 26, 29, 33, 38, 41, 100, 153, 161, 216, 233
Domfront, Château de, 206
Domes, 83–5, 92, 120–22

Domestic Buildings, 216–26, 43, 58, 100–2, 118
Donjons, *vide* Keeps,
Dörpfeld, W., 3, 4
Doune Castle, 218, 201
Dourdan, Château de, 126
Dover Castle, 79, 100, 103, 150
Drawbridges, 202
Dudley Castle, 54, 218
Durham Castle, 54, 102

ECHINUS, 18
Edessa, 45
Edward the Confessor, 52
Edward the Elder, 51
Edward I, King of England, 153, 170
Egypt, 6, 43
Elæus, 49
Elphinstone Tower, 224
Ellesmere Castle, 53
Embrasures, 204
Episcopa, 48
Escalade, 18, 142, 5
Este, Castello, 171
Etampes, Château d', 96, 133
Ethelfleda, 51
Ewias Harold Castle, 52
Exeter Castle, Gateway, 103
Eynesford Castle, 100

FALAISE, Loophole, 230
Ferns Castle, 126
Ferrara, Castello d'Este, 214
Firebrands and Fire, 14, 16, 143, 144, 145, 192
Fireplaces, 114–15, 66, 70, 72, 118–19, 120, 225
Flavius Silva, 32
Flint Castle, 155
  ,, Town, 170
Florence of Worcester, 51
Foix, Château de, 173, 114
Forebuilding, 66, 70, 76, 79
Fortunat, Bishop of Poitiers, 50
Foundations, 22, 57, 99
Framlingham Castle, 171
Fréjus, 34
Fuensaldaña, Castillo de, 224, 221

Galleries in Walls, 33, 45, 70, 76, 79, 89
Garderobes, *vide* Latrines
Garrisons, 209
Gastal, 48
Gateways and Gatehouses, 103–9, 189–202, 9, 10, 11, 23, 34, 35, 38, 46
Gaul, Cities of, 25, 33–35
Geoffrey de Vinsauf, 142, 143, 148
Gisors, Château de, 61–62, 94, 100, 122, 170, 206
Goodrich Castle, 203
Greek Fire, 49, 143, 147
Grimspound, 26
Guelma, 46
Gypsum, 22

HADRIAN's Wall, 36–37
Haidra, 46

INDEX

Halls, 100, 225-26
Hamath, 6
Hannibal, 16, 17, 19
Harlech Castle, 160, 209
Haughley Castle, 53
Helepolis, 13
Henry VIII, King of England, 231
Herod the Great, 30, 31
Herod Agrippa I, 31
Herodotus, 1, 11
Herringbone masonry, 60
Hertford Castle, 51
Hides, Raw, 14, 27, 142
Hiero II, King of Syracuse, 16
Hieron Castle, 86
Hittite Cities, 5
Hoards 205, 5, 14,
Hod Hill, 26
Hooks, 27
Houdan, Château de, 90
Houses in Towns, 10, 21, 170
Housesteads, 36-37
Huntingdon, 51

ILLITURGI, 18
Intimidation, 14, 19, 145, 146
Iron Window bars, 114
Iron Plates, 40, 142, 147
Issoudun, Château de, 125

JAFFA, or Joppa, 31
Jars, Earthenware, 22
Jerusalem, 30-31, 145
John, King of England, 148, 150
John of Gaunt, 74, 226
Josephus, 29-32, 40
Jotapha, 40
Julius Severus, 31
Justinian, 44-49

KARNAK, 6
Keeps or Donjons, Roman, 31-32
,, ,, Byzantine, 47, 86
,, ,, Shell, 57-65
,, ,, Rectangular, 66-80
,, ,, Transitional, 90-99
,, ,, Circular, 116-133
Kenilworth Castle, 74, 111, 112, 225-26
Khorsabad, 10
Kirby Muxloe Castle, 208, 230
Kitchens, 62, 72, 95, 100, 222, 225
Krak des Chevaliers, Le, 135, 202

LADDERS, 5-6, 14, 142
Lakish, 14
Lambert d'Ardes, 53, 54
Langeais, Château de, 69
Latrines, 69, 74, 83, 97, 149, 179, 182, 226-29, 231
Launceston Castle, 59, 105, 170
Laval, Donjon, 205
Lavatories, 62, 95, 97, 222
Lea Castle, 126
Lead, 22

Lead Shot, 18
Leicester Castle, 101
,, City, 35, 51
Lemsa, 48
Lewes Castle, 52, 53
Leybourn Castle, 192
Libourne, 170
Lillebonne, Château de, 128
Lilybæum, 16
Limes Germanicus, 33
Lincoln Castle, 52
Lists, 139
Livy, 12, 16, 18, 19, 20
Loches, Château de, 70, 113
London, Roman, 38
,, Tower of, 70
Longbows, 141
Longtown Castle, 98, 106, 114
Long Wall, The, 44
Loopholes or Meurtrières, 110, 74, 17, 204, 230
Louis, Dauphin of France, 150
Ludlow Castle, 103, 100, 102
Lydford Castle, 66
Lydney, 54

MACHICOLATIONS, 85, 205, 109, 125, 189-98
Madaktu, 10
Maiden Castle, Dorset, 26
Mallorca, Castello de Bellver, 165
Mangonels, 143
Maniace, Castello, 139
Manorbier Castle, 111
Mantlets, 142-43, 14, 18, 148
Mantua, Castello di Corte, 214
Marseilles, 25, 27
Martello Towers, 236
Masada, 31
Maxentius, 33
Mdauroch, 46
Medînet Habu, 9
Merlons, 204, 9
Mesopotamia, 2
Meurtrières, 74, 89, 110, 17, 230
Milecastles, 37
Mines, 14-15, 18, 28, 143, 144, 147, 149, 150
Missiles, 143, 14, 18, 144, 145
Moats, *vide* Ditches,
Monmouth Bridge, 173
Monpazier, 170
Montaner, Château de, 166
Moselle, Château on the, 50
Mottes, 52
Mount Caburn, 26
Mousa, Shetland, 25
Munich, Carlsthor Gate, 230
Munoth, Fort, 235
Musculus, 27
Mycenæ, 2

NAJAC, Château de, 126-27
Nebuchadnezzar II, 1, 11
Nether Hall, Stairway, 208
Newark-on-Trent, Gatehouse, 105
Newark, Selkirkshire, 224

INDEX

New Carthage, 19
Newcastle, Keep, 76
Nicæa, 42
,, Siege of, 144
Nîmes, 34
Nimrûd, 9

Oakham Hall, 101
Old Basing, 52
Old Cairo, 43
Old Sarum, 62
Olive Wood, 22
Orford Castle, 94
Ortenberg Castle, 125
Orthez, Bridge, 172
Oubliettes, 83
Outworks, 22, 116, 153
Ovens, 98, 74, 95, 122
Oillets, 216

Palestine, 30–33, 145–48, 40
Palisades, 22, 16, 26, 33, 52–53, 54–57
Palus, 225
Pantry, 226
Parapets, 204–6, 9, 14, 24, 33, 45, 46, 110
Paris, Siege of, 141
,, Bastille, 215
,, Château de Vincennes, 218
Parlour, 222
Parthenay, 190
Pele-towers, 225
Pembroke Castle, 119, 112, 189
Pendennis Castle, 235
Pendragon Castle, 179
Penthouses, 27, 142
Périgueux, 230
Petrariæ, 143, 15, 18
Pevensey Castle, 39–40, 24 51
Pfeffengen, Château de, 62
Philip of Macedonia, 14
Philip V of Macedonia, 16, 18
Philip II, King of France, 148
Philo of Byzantium, 12, 21
Pigeon Holes, 98, 186
Pierrefonds, Château de, 210, 218
Pistes, Edict of, 50
Pit, 182, *vide* Prison,
Pleshey, 170
Plans of Fortresses, 21, 29, 116, 153, 232
Plinths, 202, 59, 72, 83, 117, 121, 135, 164, 214
Polybius, 16–20, 27–28
Pomerium, 170
Pompei, 23–24
Porch, 76, 94, 225
Porchester Castle, 39, 40
Portcullises, 17, 189, 30, 34, 71, 76, 180, 216
Posterns, 22, 2, 4, 48, 62, 66, 103, 122, 140, 153
Prætorium, 29, 37
Procopius, 42, 141
Prior Laurence, 54
Prisons, 226, 83–85, 59, 103
Provins, Tour de César, 92
Prows, or Spurs, 117–19, 100, 125, 190, 203

Queenborough Castle, 166

Ramesseum, 6, 5
Ramp, 1, 2, 3, 210
Rampart, 38
Ravelin, 213
Reculver, 39
Restormel Castle, 60
Rhodes, 12, 169
Richard's Castle, 52, 54
Richard, Earl of Cornwall, 59
Richborough Castle, 39
Reichenberg, 99
Richmond Castle, 101, 102, 100
Roche Guyon, La, 118, 122
Rochester Castle, 75, 115
Roger, Bishop of Salisbury, 62, 101
Rollo, 141
Roman Camps, 29, 16, 36–37
Rome, Castra Prætoria, 30
,, Walls of, 33
,, Siege of, 141
Roofs, 66, 74, 75, 76, 94, 179, 198
Roumeli Hissar, 83
Roumeli Kavak, 86
Rouen, Tour Jeanne d'Arc, 122

St. Andrew's Castle, Mine, 150
St. Honorat, Château, 80
St. Mawes Castle, 235
Saladin, 89, 143
Salapia, 17
Salignano, Fort, 230
Sallyport, 22, 140, *vide* Postern
Sambucæ, 16
Sandgate Castle, 234
Sapping, 14, 90, 142, 144, 202
Sarzanello, Castello di, 163, 206
Saxon Shore Forts, 39
Scipio, 19
Scorpians, 143
"Screens", The, 225 222, 216,
Senlis, 35
Shell Keeps, 57, 65
Sherborne Castle, 101, 104–5, 100
Shields, 5, 14, 27, 149
Ships, 16, 19, 144
Shutters, 204
Sieges, 144–150, 16
Siege Engines, 142–143, 14, 18, 27, 146, 147
Siege Towers, 142, 14, 18, 27, 145, 147
Signal Towers, 33, 37, 46
Sinjerli, 6
Sirynx, 16
Skenfrith Castle, 99, 100, 114, 52, 57
Slings, 14
Smailholm Tower, 225
Smederevo, 171
Solar, 226, 70
Spurs, 203, 100, *vide* Prows
Stairways, 207, 43, 58, 59, 66, 79, 97
Stephen, King of England, 57, 142
Stone bosses on face of Towers, 22, 65
Strabo, 1

Streets, 10, 29, 170
Swords, 14
Syracuse, 16–19, 139

TACITUS, 39
Tamworth Castle, 60
Tantallon Castle, 213
Taranto, 19
Tarragona, 24
Tattershall Castle, 222
Temenos at Ur, 11
Tête-de-Pont, 172, 173
Thebes, 6
Thélepte, 46
Theodorus Silentiarius, 48
Theodosiopolis, 45
Threave Castle, 182
Tiberius, 30
Tickhill Castle, 54
Tiffauges, Château de, 206
Tiglath-Pilesar III, 14
Timber bonding, 22, 25, 62, 121–22, 185
Timgad, Citadel, 47
Tiryns, 3
Titus, 40
Topcliffe, 54
Torches, 14, 143, *vide* Firebrands,
Tortoise, 27
Totnes Castle, 58, 170
Tournai, Pont des Trous, 172
Towcester, 51
Towers, 21, 22, 100, 126, 173, 224
Towns, 170, 10, 21
Trebuchet, 143
Trees in Ancient City, 10
Trematon Castle, 60
Treves, 35
Triangular Castles, 163
Troy, 4
Troyenstein Castle, 213–14
Tunisia, 46
Turks, 145, 147, 230

Tyre, William of, 144–46

UR of the Chaldees, 11
Uronarti, 6
Ursino, Castello, 139
Utrecht, Psalter, 51

VALLUM, 36
Verona, 172, 171
Vespasian, 33, 40
Vegetius, 85, 192
Villafranca, Castello, 171
Villandraut, Château de, 163
Villard de Honnecourt, 143
Vinæ, 27
Vincennes, 218
Viollet-le-Duc, 84, 143, 229
Visby, 171, 203
Vitruvius, 22, 14

WADI Halfa, 6
Walls, *vide* Curtain walls,
Walls, Arcaded, 12, 46, 85, 86, 89
Wall walks, 204, 21, 22, 33, 45, 46, 110, 133, 224 157, 159, 163, 198
Walmer Castle, 232
Ward or Bailey, 52, 100, 116, 135, 149
Warkworth Castle, 218, 116, 135, 149
Warwick Castle, 177, 198, 226
Water Supply 31, 45, 79
Wells, 69, 72, 74, 79, 94, 100–1, 182
William the Conqueror, 52
William of Poitiers, 53
Winchelsea, 170
Windows, 112
Winklebury Castle, 26
Witches, 146

YETTS, 198
York, City, 35, 51
 ,, Multangular Tower, 40
 ,, Clifford's Tower, 133

DATE DUE			
GAYLORD			PRINTED IN U.S.A.

21962